IMAGES OF ASIA
Series Adviser: SYLVIA FRASER-LU

Songbirds in Singapore

Titles in the series

At the Chinese Table
T. C. LAI

Balinese Paintings (2nd ed.)
A. A. M. DJELANTIK

The Birds of Java and Bali
DEREK HOLMES and
STEPHEN NASH

The Birds of Sumatra and
Kalimantan
DEREK HOLMES and
STEPHEN NASH

The Chinese House: Craft,
Symbol, and the Folk Tradition
RONALD G. KNAPP

Chinese Jade
JOAN HARTMAN-GOLDSMITH

Early Maps of South-East Asia
(2nd ed.)
R. T. FELL

Folk Pottery in South-East Asia
DAWN F. ROONEY

Fruits of South-East Asia: Facts
and Folklore
JACQUELINE M. PIPER

A Garden of Eden: Plant Life in
South-East Asia
WENDY VEEVERS-CARTER

The House in South-East Asia
JACQUES DUMARÇAY

Images of the Buddha in Thailand
DOROTHY H. FICKLE

Indonesian Batik: Processes,
Patterns and Places
SYLVIA FRASER-LU

Japanese Cinema: An Introduction
DONALD RICHIE

The Kris: Mystic Weapon of the
Malay World (2nd ed.)
EDWARD FREY

Life in the Javanese Kraton
AART VAN BEEK

Macau
CESAR GUILLEN-NUÑEZ

Mandarin Squares: Mandarins
and their Insignia
VALERY M. GARRETT

Musical Instruments of
South-East Asia
ERIC TAYLOR

Old Bangkok
MICHAEL SMITHIES

Old Manila
RAMÓN MA. ZARAGOZA

Sarawak Crafts: Methods, Materials,
and Motifs
HEIDI MUNAN

Silverware of South-East Asia
SYLVIA FRASER-LU

Traditional Chinese Clothing in
Hong Kong and South China
1840–1980
VALERY M. GARRETT

Songbirds in Singapore
The Growth of a Pastime

LESLEY LAYTON

SINGAPORE
OXFORD UNIVERSITY PRESS
OXFORD NEW YORK
1991

Oxford University Press

*Oxford New York Toronto
Delhi Bombay Calcutta Madras Karachi
Petaling Jaya Singapore Hong Kong Tokyo
Nairobi Dar es Salaam Cape Town
Melbourne Auckland
and associated companies in
Berlin Ibadan*

Oxford is a trade mark of Oxford University Press

© *Oxford University Press Pte. Ltd. 1991*

*Published in the United States by
Oxford University Press, Inc., New York*

All rights reserved. No part of this publication may be reproduced,
stored in a retrieval system, or transmitted, in any form or by any means,
electronic, mechanical, photocopying, recording or otherwise,
without the prior permission of Oxford University Press

ISBN 0 19 588999 1

British Library Cataloguing in Publication Data

*Layton, Lesley
Songbirds in Singapore: the growth of a pastime.
1. Singapore. Pets
I. Title   II. Series
636.0887*

ISBN 0-19-588999-1

Library of Congress Cataloging-in-Publication Data

*Layton, Lesley, 1954–
Songbirds in Singapore: the growth of a pastime / Lesley Layton.
p.   cm. — (Images of Asia)
Includes bibliographical references and index.
ISBN 0-19-588999-1:
1. Cage birds—Singapore. 2. Cage birds—Social aspects—
Singapore. 3. Bird-song. I. Title. II. Series.
SF461.43.S55L38   1991
636.6'8'095957—dc20
91-13698
CIP*

*Printed in Singapore by Kim Hup Lee Printing Co. Pte. Ltd.
Published by Oxford University Press Pte. Ltd.,
Unit 221, Ubi Avenue 4, Singapore 1440*

To my mother and father

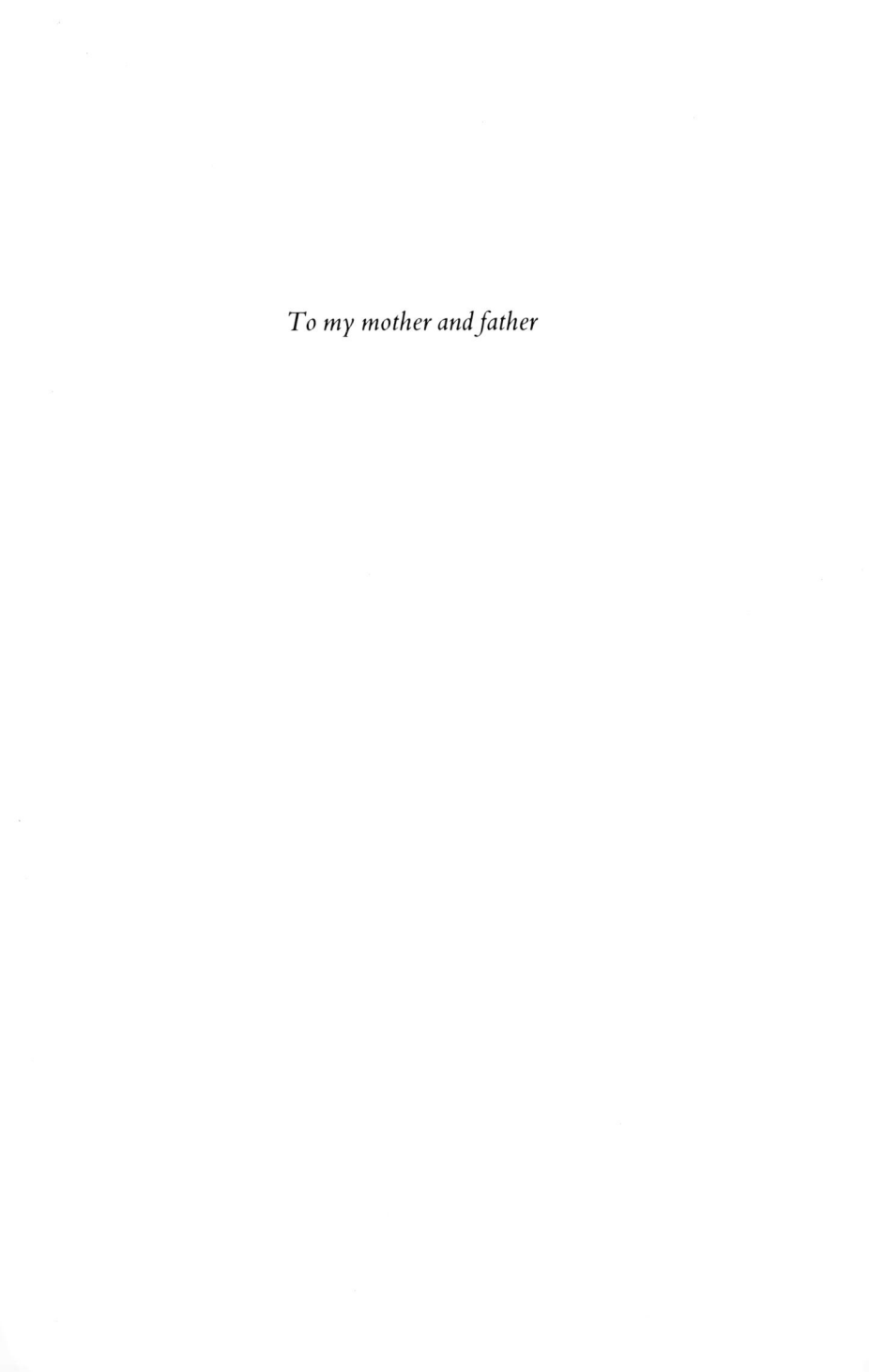

Preface

MUCH has been discovered and recorded about the behaviour and habits of birds both in nature and in captivity, but very little concerning their relationship with man throughout the ages and their importance in his life. In Asia, still less has been explored and written about this, although bird-keeping in this part of the world is very widespread. Tracing its history must therefore be based on memories, on habits passed on from one bird-keeper to another, on surviving artefacts, and to a large extent on conjecture.

This book sets out to describe the practice of songbird-keeping in Singapore as it has evolved through the interrelationship of the peoples of China, India, and the Malay Archipelago—and to awaken an interest in a facet of life which, sadly, seems to have been long considered unworthy of close scrutiny. The intention is not to condone the imprisonment of birds in small cages, nor to encourage participation in the hobby, but to trace its development over time and discover some of the meaning that lies within it.

While doing so, it is very important to understand that attitudes to birds and bird-keeping in Asia are quite different from those in the West. Among the Chinese, for example, songbirds are sincerely thought to be better off in cages and are given a great deal of time and attention; therefore, Western feelings of sympathy towards them may be somewhat misplaced.

The introduction to the book provides background information, showing how birds permeated the lives and culture of the three main races—Chinese, Malay, and Indian—that made up the society of Singapore from its early days. The book then describes how the practice of bird-keeping, imported from such diverse corners of the region, came to grow and blossom in Singapore. It examines the influences of one race upon the other and the reasons why songbirds became particularly favoured.

The most sought-after songsters are then described and how they are chosen, cared for, tamed, trained, and entered in song contests. This is followed by a description of their cages and cage accessories, a brief overview of the modern practice of breeding songsters, and finally, a consensus of feeling about the future of the pastime, or 'playing birds', as it is affectionately known.

The text contains early illustrations of traditional practices in bird-keeping, as well as recent photographs showing many facets of the local passion for songbirds. As this book is mainly about 'playing birds' in Singapore, local terms are used in the text, and as soon as each of the songbirds has been formally introduced in Chapter 4, they are referred to as they would be locally.

For making the book possible, thanks are due to the following: Tan Beng Chong, Education Officer, and Khu Siak Hui, Keeper of the South East Asian Birds' enclosure at the Jurong Bird Park; C. P. Chew and Amin Rahmat, for demonstrating the cage-making process in their homes; Tony Wong, collector of cages and bird-feeders; Steven Lew and Chia Eng Seng, judges; David De Souza and Lin Heau Dong, breeders of Shamas; S. Wee and John Yim for sharing their memories of 'old times', and to Ali Mohamed Yusoff for his invaluable information on the Zebra Dove.

Finally, special thanks are due to Michael Sweet of Antiques of the Orient for the early illustrations; to

PREFACE

Guus van Bladel, David De Souza, and John Yim for their advice and support, and to Judith Balmer for the enthusiasm and care that she has put into the photography.

Singapore   LESLEY LAYTON
January 1991

Contents

Preface — vii

1. Introduction — 1
2. The Growth of Bird-keeping in Singapore — 9
3. Why a Songbird? — 19
4. The Most Popular Songbirds — 26
5. Beginning with a Songbird — 34
6. Songbird Care and Training — 44
7. Bird-singing as a Sport — 55
8. Cages and Accessories — 66
9. The Future of the Songbird Tradition — 77

Select Bibliography — 86
Index — 87

I
Introduction

BIRDS have been kept all over Asia for thousands of years, not only for their beauty and their song, but for amusement and profit. They have also played another, more significant, part in people's lives: as symbols within two of the most prominent belief systems, Hinduism and Buddhism. The same birds, common to both, appear in the folk myths and superstitions that surround them, and are used repeatedly in art forms.

The Chinese have always been very fond of birds and evidence of this can be traced back to their earliest ancestors who painted birds on pottery and carved them in jade. Images also appeared in pictographs on oracle bones[1] and in the decoration of the first bronzes. A bronze birdcage from the Warring States period (481–221 BC), now in the Asia Art Museum in San Francisco, is early evidence of a desire to keep birds in cages.

Birds were frequently included in the literature and poetry of China, and the painting of birds was elevated to a class of its own. Captive birds, though used as models for paintings, were rarely painted in cages. When a cage was added, it usually appeared as an extrinsic detail—a gift brought in tribute or part of a celebration taking place (Plate 1). Such a subject was so commonplace in China that depictions were considered to be only of interest to foreigners or overseas Chinese.

---

[1]Animals' shoulder blades were used at the Shang court in China (1650–1027 BC) for looking into the future. The king would question and the diviner would apply a heated rod to the bone and interpret the answer from the pattern of cracks on its surface.

1. 'Celebrating the New Year' by Ting Kuan-p'eng, court artist of the Ch'ing dynasty (1644–1911). In this silk scroll, a birdseller is trying to attract the attention of people who have come to watch a performance. (Courtesy of the National Palace Museum Collection, Taipei, Taiwan.)

2. Moulded plaster eagles were once common on gateways in Singapore, but this unusual house shows them as roof finials. (Lesley Layton)

In Singapore, where Chinese representation dominates, birds are one of the most common symbols to appear as central motifs on silks and ceramics, in bronze or stone, and as architectural reliefs (Plate 2). Examples of these symbols include the crane, vehicle of the gods and symbol of longevity; the goose and duck, representing close and harmonious relationships; and the peacock and pheasant, possibly for their resemblance to the mythical phoenix, once associated with the empress of China.

The phoenix was also of special significance to the Straits Chinese. These people were a separate community born of the marriage between early Chinese traders and Malay women in the Straits Settlements of Malacca, Penang, and Singapore. Although they have now been largely absorbed by the indigenous population, the phoenix motif survives on their silverware, ceramics (Colour Plate 1), and embroideries. Other evidence of Chinese designs tempered by Malay influence is found on the surviving homes of the Straits Chinese, with their characteristic *pintu pagar* (saloon-type half doors) (Colour Plate 2), and coloured ceramic tiles, often decorated with birds.

The release of birds from captivity, an ancient Buddhist method of gaining merit, was already associated with freedom from oppression when Buddhism reached China from India. It was an act that also came to be synonymous with celebration. When the Empress Dowager Tz'u-hsi produced an heir to the throne, she ordered that all the caged birds of China be released. Mrs E. F. Howell, writing on native superstitions in 1923, found that in Singapore 'this custom is known as Fong Sang: a person having the idea that by releasing these living creatures they will secure rewards in the world to come' (Wise, 1985: 203). More recently, a Chinese man in Singapore celebrated the birth of his firstborn son by releasing

INTRODUCTION

10,000 Java Sparrows from a temple, to give the boy a good start in life.

Popular among trained birds in China were carrier pigeons and parrots. It was customary for pigeons to be released during processions of Ch'ing emperors, often with whistles attached to their tails to provide musical accompaniment. Parrots, considered 'holy birds' because of their ability to talk, were associated with the palace concubines on whom they were said to spy, later repeating their conversations to the eunuchs.

Cock-fighting, learned from the Malays, was a popular sport among Mandarins, while the common people had a greater interest in quail-fighting (Plate 3), especially in northern China. In 1859, Revd G. N. Wright, a pioneer missionary, witnessed how 'The victor is put up for sale or raffle and the eagerness to become his master is demonstrated by the enormous sums staked or paid down for him' (Tate, 1988: 97). The Chinese did not always train birds for amusement. Important to the livelihood of many country people was the cormorant, a sea bird which was trained to catch fish and return with them to the fisherman's boat.

Songbirds were prized among senior officials and eunuchs of the emperor's court in Peking (now Beijing) during the Ch'ing dynasty (AD 1644–1911), though by the nineteenth century most gentlemen would have owned one themselves. There is an old tale about a group of envoys who attended a meeting accompanied by their birds; but the birds became so noisy that the meeting had to be adjourned until they had been appropriately soothed and attended to. This excessive pampering is an integral part of songbird care that persists to the present day.

The abundance of old, beautifully made cages and traps found in the Malay Archipelago is evidence of the

3. Quail-fighting in old China, from a steel engraving after a drawing by Thomas Allom, from Revd Wright's *China Illustrated*, London, 1843. (Courtesy of Antiques of the Orient, Singapore.)

long-standing importance of birds, mainly doves and quails. Birds such as these were said to contain spirits of the dead, an animistic belief which accounts for the Malay conviction that a dove had the power to bring luck and the owl was a messenger of death. Hornbills were kept as pets in Iban longhouses and are common in their mythology. Wild birds were said to carry the wisdom of the ancestors and the Iban people would look to these 'omen birds' for guidance.

In 1872, the *Illustrated London News* stated that 'The Malay peninsula and the islands around it are celebrated for their fauna ... cockatoos and parrots of every size and colour are to be found' (Tate, 1989: 23). The same source revealed that at that time it was 'common for a boy, when he has finished his career as a diver, to make a new start, and become a birdseller'. A little later, in 1883, Isabella Bird, on her extensive travels in the region, found the Malays to be 'passionately fond of pets' ... to have 'much skill in taming birds' and their children 'encouraged to invite the confidence of birds ... rather than torment them' (Bird, 1883: 300).

The Indian mythological bird Garuda, symbol of power and strength, figures prominently in ancient Malay and Indonesian literature, art, and dance drama. Its ancient Hindu connotations were lost with the rise of Islam but folk versions survive. With Islam, both awe and respect grew for the Zebra Dove because of its inherent ability to reproduce sounds from the alphabet of the holy Koran.

In India, doves and talking birds have always been revered. A favourite pastime of the Mogul emperors of Delhi in the sixteenth to nineteenth centuries, the ancient game of pigeon-flying is now popular among common people, while parrot-keeping was widespread long before

Alexander the Great brought these birds to Europe. Indian literature and folk myths include a wealth of tales about birds, especially parrots, peacocks, and water birds. A painting from the Rajput period (seventh to eighteenth centuries) shows a parrot being offered rubies, which it mistakes for pomegranate seeds, to discourage it from repeating a romantic interlude it has overhead. These same birds are associated with gods or aesthetic achievements. Saraswathi, the goddess of learning, is likened to a swan. In Singapore, peacock feathers are an important decoration in the festival of Thaipusam, to honour Lord Subramaniam whose vehicle is a peacock.

Modern Singapore continues to demonstrate its great love of birds by featuring them on banknotes and postage stamps (Colour Plate 3) and in television documentaries and decorative displays, while stylized versions are common as corporate logos. While bird-singing contests are still in their infancy, bird-keeping itself is a pastime with a long history, and one which the three races that settled here were familiar with, and receptive to. With a rebirth of their interest in a new land, each of the races inevitably began with their own traditional practices and gradually borrowed those of others. In the ensuing process of growth and development, the hobby has acquired some characteristics uniquely Singaporean, while at the same time maintaining close links with the other countries of the region that share the same passion.

2
The Growth of Bird-keeping in Singapore

IN its formative years, Singapore was a stopping-off point for trade in all things exotic, and birds were no exception. The 1872 December issue of the *Illustrated London News* reveals it to be 'one of the best places for buying birds of that kind', and P & O steamers arriving at the island in the nineteenth century would find a Malay birdseller at the landing-stage, hoping to make a sale (Tate, 1989: 23). The same issue goes on to explain that 'The perch on which he [the birdseller] holds the bird is a piece of bamboo; the hollow place at the top holds food, and a bit of wood is stuck in at the side, upon which the bird sits' (Plate 4).

Annie Brassey, sailing round the world with her family in 1881, came across a bird market in Singapore, finding it 'a curious place.... The whole street is alive with birds in baskets, cages and coops, or tied by the leg and thrown down anyhow ... they are all very tame and very cheap' (Wise, 1985: 102). The bird trade was originally centred around the Rochor River, an important waterway and already the site of a settled population when Sir Thomas Stamford Raffles arrived in 1819. Once the town of Singapore was established, this area became the island's wholesale trading centre, and retail market-place (Plate 5).

Probably the most alarming traders from the region were the fierce Bugis in their prahus, because of their reputation for indulging in piracy, though the screams from their cargo of exotic birds never failed to attract

4. 'Malay Birdseller at Singapore', from the *Illustrated London News*, 7 December 1872. (Courtesy of Antiques of the Orient, Singapore.)

5. Rare etching of Rochor River, showing early settlement and a trading vessel, by F. Jagor, from his *Singapore–Malacca–Java Reiseskizzen*, Berlin, 1866. (Courtesy of Antiques of the Orient, Singapore.)

people to the beach from where they did their trading. Mrs Brassey remembers witnessing the less colourful traders that brought their birds upriver: 'Towards the end of the south-west monsoon, little native open boats arrive from the islands 500 to 1,500 miles to the south-eastward of Singapore. Each has one little tripod mast. The whole family live on board. The sides of the boat cannot be seen for the multitudes of cockatoos, parrots, parakeets, and birds of all sorts, fastened on little perches, with very short strings attached to them.... With this cargo they creep from island to island, and from creek to creek, before the monsoon, till they reach their destination. They stay a month or six weeks.... They then go back with the north-east monsoon, selling their goods at the various islands on their homeward route. There are many Dutch ports nearer than Singapore, but they are over-regulated, and preference is given to the free English port ...' (Wise, 1985: 102).

Roland Braddell, writing in the early twentieth century, states that 'The Singapore animal trade was started in 1880 by a Malay named Haji Marip who carried on the business until his death in 1915. Many followed his example, and in Rochor road today [1934] there are several Chinese dealers' (Braddell, 1934: 123). These probably set up their businesses in the *lo ma pan jiang*,[1] a row of six or seven shophouses in Rochor Road, which were the first 'pet shops'. Two of the shops, namely Jin Seng and Ang Soon Huat, went on to build a reputation for their birds that lasted well into the 1980s when the latter finally closed (see Plate 9). The Jin Seng bird shop is still run by descendants of the original owner, although it has changed its name and location.

[1] The Chinese name for the Malay term *rumah panjang* or longhouse.

THE GROWTH OF BIRD-KEEPING IN SINGAPORE

In the years leading up to the war, there was not much time for pet-keeping, especially in the burgeoning city. In outlying areas, however, pigs and poultry were farmed, and wild birds such as Magpie Robins, were attracted there by the food. Toa Payoh, meaning 'great swamp', was a typical example of an outlying area where people were few and wild birds were plentiful. A retired man, remembering his childhood, claimed that 'Bishan park, too, was a haven for wild birds, and near the Pierce Reservoir there was a big stretch of land where there were many young birds for the taking. People were not as territorial as they are now, and a boy could help himself to fruit from the trees and a young bird or two. In those days', he said, 'birds were let out of their cages to fly free, and return.'

Those not taken from the wild were bought from animal markets, such as the one in Trengganu Street in Chinatown, or the shops in Rochor Road. Quite a varying collection of creatures were bought and sold, mainly for medicinal purposes, and to supply the zoos in the area. Braddell mentions the 'Singapore Zoo at Ponggol' where there was a 'magnificent collection of birds' and large private collections were not uncommon.

Birds were popular as pets among the British armed forces and their families in the 1940s. In 1945, they began to club together and exchange information on bird-keeping. This led to the formation of the Singapore Caged Birds Society which met each month at the homes of its various members. Local people were invited periodically to attend their bird shows. An elderly Singaporean remembered these as a family affair: 'Children brought their pets and old people pushed along their caged birds in prams.' The birds displayed were judged for their appearance, health, and tameness. Parrot species such as

Budgerigars, Lorikeets, Lesser Sulphur-crested Cockatoos, and Hanging Parakeets were the most common but there were also songbirds such as Canaries and Magpie Robins.

Bird ownership did not really begin in earnest until well after the Second World War when people renewed their pre-war interests and bird shops began to fill up again with songsters from Malaya and Lorikeets, Cockatoos, and Hill Mynahs from Indonesia. An interest in baby mynahs may have been one of the reasons for the spread of these birds in Singapore, a species not seen before the war. Canaries were imported from China just before the arrival of the Japanese in 1942, but did not really become popular until much later when air travel brought them from Europe.

The practice of trapping songbirds in Malaya for sale in Singapore began in the early 1950s, the birds travelling over the causeway in the frequent trucks carrying consignments of fruit. By the 1960s they were being trapped over a wide area, from Kota Tinggi to the tip of Johore. At first, the attraction was Zebra Doves and then Oriental White-eyes, Red-whiskered Bulbuls, and White-rumped Shamas. China Thrushes were seen only occasionally in Singapore until the 1960s when they were imported from China. Before this, they were brought in by seamen who would congregate around Sago and Smith Streets, the area of the letter-writers in Chinatown, and exchange their birds for other goods at the market nearby.

From the late 1950s, a greater number of people had started to keep birds, and it is interesting to note that most began by keeping those that were traditionally kept in their homelands. Even when other species were collected, certain types of birds became associated with one or other of the ethnic groups. The Chinese kept Magpie Robins and Oriental White-eyes. To these were added the

Skylark, Mongolian Lark, and China Thrush. The Malays kept Red-whiskered Bulbuls and various species of dove, including the Spotted Dove and the Zebra Dove. The Javanese community kept Hanging Parakeets and Pink-headed Doves.

The Straits Chinese kept birds that both the Chinese and the Malays enjoyed, as well as Canaries, Sulphur-crested Cockatoos, and Pekin Robins. The Eurasians and the English-educated kept Canaries, Red-whiskered Bulbuls, and mynahs. The Indians kept mynahs, Java Sparrows, several types of dove, and Ring-necked Parakeets. The demarcation was due to the various communities living segregated lives and keeping their knowledge about bird-keeping to themselves.

It was only in the 1960s and 1970s, as people were moved from their country kampongs into new flats in the city (Plates 6 and 7), that communities began to intermix and enormous changes came about in bird-keeping practices. At first, there was a lull in bird-keeping because it was felt that high-rise flats would be too small. This was not found to be so, and when meeting other people carrying cages of birds made it easier to make friends, the hobby began to grow again. From the early 1960s, there was also a rise in the number and species of birds available at the handful of shops in Rochor Road, and a corresponding rise in the number of bird shops themselves.

By this time, a number of songbirds were owned throughout Singapore, mostly by elderly men. Although little attention was paid to their birds' song, other than pleasure in the sound, these men began to gather together to display their birds and to discuss how to improve the health and appearance of their birds. Lack of information at that time, especially about diet, prevented bird-keeping from growing even faster than it did.

6. Malay kampong house in the 1950s, showing a dove cage hanging from the eaves (top left). (Courtesy of the National Archives, Singapore.)

7. Modern Housing Development Board flats showing a variety of birdcages hanging above the walkway. (Judith Balmer)

From the 1970s, songbird-keeping began to escalate, with the quality of song becoming the focus of attention, and song contests gaining in importance in many parts of the region. By the 1980s, Singapore's bird-singing contests had acquired a large following and the phenomenon was added to the country's list of tourist attractions. For the local people, surprised by the sudden outside interest in the activity, it was simply a harmless and practical answer to a more restricted lifestyle, and an ancient pleasure.

3
Why a Songbird?

BIRDS of all kinds were looked upon with affection, but it was songbirds above all others that made their mark on Singapore. This was partly due to chance, but also to shared beliefs about these birds, to an environment that lent itself to keeping songbirds, and to circumstances that allowed it to grow from a pastime into a competitive sport.

The first Chinese immigrants who came to the island in search of work or trade would not have had the means to keep songbirds. Much later, when times were not as hard, they trapped them for their sweet voices or their fighting ability. Richer Chinese arriving from islands in the archipelago may have brought their songbirds with them.

Meanwhile, there were already 'singing' doves kept by the Malays in kampongs along the coast and inland rivers. Any links between these and other Malay communities in the region would have perpetuated their bird-keeping practices. Indians were the least likely of the three races to have kept songbirds, preferring talking birds such as parrots and mynahs, but some species of dove were enjoyed and this would later lead to an interest in Singapore's most expensive songbird, the Zebra Dove.

Among all three races, birds are associated in some way with the power to bestow luck or misfortune. In Chinatown there are fortune-telling cockatoos, as there are parakeets in Serangoon Road, the 'Little India' of Singapore (Colour Plate 4). The customer's name and date of birth are called out and then the bird picks out a card from a pack with its beak, or in some cases a numbered bamboo stick, which is then read by the astrologer.

For the Malays, the Zebra Dove has long been considered a bearer of health and prosperity, even though a Central Javanese legend says it is a man that has been cursed. Legends that tell both good and bad of the bird persist, but they are believed less and less as time goes by. Some Chinese Singaporeans also consider the Zebra Dove to be a magical bird. It is said to be lucky for certain owners, but can equally be a harbinger of misfortune, and should be given up by those who do not profit by it. One man reported that a series of accidents prevented him from catching a wild one, until eventually he gave up the chase. Another said that the bird had turned into a snake during the night.

If the bird suits its owner, it is said to look after him and protect his family. A Chinese owner described an actual event which persuaded him of the truth of this saying: 'One day a row of shophouses caught fire and my home was in danger of burning. My bird was in its cage, which was hanging from the roof at the back of house. I stood nearby, hosing the house with water in the hope that this would prevent it from catching fire. Suddenly, a great wave of flame came over the wall behind me. The bird called out and the wave of fire doubled back and sank behind the wall. It was not the wind because the flames were sheltered by the house.' Word of this unusual occurrence soon spread and a medium asked him for the droppings of this particular bird for medicinal purposes. Wary of the extent of the bird's powers, he refused.

An ancient strain of the same species, referred to as the

Majapahit[1] bird, is thought to have been a charmed bird. Sometimes it is called the 'Datuk Burung', or 'Grandfather Bird', a title of reverence stemming from old Malay/Indonesian animistic beliefs. The Zebra Dove is also said to drink very little water, and leave no smell in the cage when it dies.

Purchasers of the Zebra Dove are known to count the scales on its legs very carefully, because of a belief that a certain number signifies a lucky bird. Just how many is difficult to determine. Bucknill and Chasen, writing in the 1920s, admit to the same problem: 'We have been singularly unsuccessful in getting Malays to agree on this point and although we picked out the older men to interrogate they seemed curiously at variance as to the amount of the lucky number' (Bucknill and Chasen, 1927: 58).

This practice is not taken as seriously as in the past, and the number seems to depend more on what the individual decides will be lucky for him. However, thirty-six scales are said to be definitely unlucky and suggest a cursed bird, while thirty-three scales indicate a special bird with a good 'kong' sound. In the past, little attention was paid to this aspect of the bird's song, but now it is the most important attribute of the Zebra Dove. The sound itself has become associated with luck because a particularly powerful 'kong' may determine a winner in a song contest.

The Magpie Robin was once considered unlucky because a black-and-white plumage was associated with mourning (see Colour Plate 11). Those who did not believe such tales found that the Magpie Robin could

---

[1] A thirteenth to fourteenth-century empire with its capital in East Java, which covered most of Indonesia and included Singapore.

fetch a large sum of money even then, because it was both a good singer and a good fighter. The China Thrush, another fighter that eventually became very popular as a songbird, was at first shunned by some Chinese. This was because its natural call was similar in sound to a dialect word that, repeated, was thought to invite disaster.

Gambling was a common pastime in old Singapore. The Chinese enjoyed laying bets on fighting birds, as did the Malays and Indians. Cock- and quail-fighting were common, but owners of songbirds with 'enough fire' to be fought would often agree on a wager. The cages would be placed side by side, partially covered, and the birds allowed to challenge each other. Then, as soon as each bird was prepared to lunge at its adversary, the cage door would be opened and people would crowd around, betting.

The birds would attack each other, pecking fiercely, one pinning the other down with its claws. Its rival would lie motionless for a while, but once released would counter-attack. China Thrushes would fight only briefly, while it would not be unusual for Magpie Robins to take an hour or more to establish a winner. The battles were intense and often violent, but injuries were rarely severe because, once discredited, the loser would retreat in shame.

After these fights were made illegal towards the end of 1965, changes had to be made in the way birds were appraised. Meanwhile, an interest had developed in birds with strong voices that could make a variety of sounds, and they fetched higher prices as a result. The idea of concentrating on certain qualities in the birds thus grew in popularity, and bets began to be laid on the quality and variety of their song, the length of time they could sing, and the presence of certain body movements.

Friends looking on would decide which birds would

qualify, and outsiders would join in to select the best bird. The winner would be the one that the most people favoured. Enthusiasts would club together and buy a drink or a meal for the owner of the winning bird. This practice was so much enjoyed that it began to be repeated on a regular basis, with the group of interested people growing all the time.

In 1963, the Kelab Burung Singapura (Singapore Bird Club), a local equivalent of the earlier, largely expatriate Singapore Caged Birds Society, was started, which was later to stage the first formal bird-singing contest. Meanwhile, meetings and note-swapping were held at various venues, such as the three main bird shops in Rochor Road, the area of the porksellers along South Bridge Road, in Petain Road, which was—and still is—a common meeting-place for displaying birds (Colour Plate 5), or in parkland with trees from which the cages could be hung. Birds of each type tended to be brought to specific places. The Oriental White-eyes, for example, were displayed in the grounds of St Andrew's Cathedral, near the fruit and flower trees to which they were most attracted.

During the late 1950s, the Chinese community began to favour certain coffee-shops as the place to meet for a cup of coffee or a light breakfast, and to exchange notes on bird-keeping. The habit was an old one, popular since the beginning of the century among rickshaw pullers who met at coffee-shops, and Indian workers who gathered at the toddy-shops in Rochor Canal Road, glad of a short respite from unforgiving work and congested homes.

The Ang Siang Hill coffee-shop in the heart of Chinatown was very popular on weekday evenings and Sunday mornings with songbird owners, probably because it had once been the best place to search for opponents for bird fights. In the early 1970s, the coffee-shop moved out of

Chinatown to Tiong Bahru Road, and became a gathering place mainly for Oriental White-eyes, which have never been fighting birds (Colour Plate 6).

Although the shop is officially called Wah Heng, it is simply known as 'the Tiong Bahru coffee-shop', and its reputation has made it a popular venue for tourists. Above the coffee tables, a series of horizontal metal poles have been constructed, from which hang numbered wires for attaching cages. The idea was improvised from the earlier custom of suspending Merbok cages on vertical numbered poles. A small group of men gather here each day, but Sunday mornings are the shop's busiest time, when people travel from all corners of the island just to display their birds here.

Both Malays and Chinese meet, with their birds, at coffee-shops all over Singapore before dawn or in the evening after work, when birds are most vocal. Here they sit whistling to the birds, comparing notes, listening to other birds, and discovering their own bird's position in the street market. If they hear another bird produce a few unusual notes, they will hang theirs next to it, in the hope it will add this new sound to its repertoire. For the birds, the outing simulates a contest environment, an opportunity to grow used to singing for longer periods of time among crowds of people and birds, as well as a way to improve the variety of their song.

The widespread interest in this hobby did not go unnoticed by the Singapore Government who, from the mid-1970s, set out to provide every large housing estate with a shop selling birds. The intention was to foster an empathy for live creatures among residents, though one young man playfully suggested that it was also because 'while we are talking about the birds, we are not hatching up a riot or complaining about the system'. Certainly, it has proved to be a peaceful hobby which cuts across language, race, and rank.

All new housing estates and parks have been given special bird corners equipped with poles and wires to hang birdcages, usually located beside or close to coffee-shops, and the most popular ones, such as those at Ang Mo Kio and Bedok, are very crowded. Bird-singing clubs have been established in most residential areas by members of the community. Almost every weekend, song contests are arranged and the money raised is used to finance a variety of community activities.

Bukit Batok, a predominantly Malay area of Singapore, provides a striking example of how the hobby can grow within one residential district. Here, keeping Zebra Doves has become so widespread that three or four residents out of every ten own one. The neighbourhood gardens and void decks—large, open recreation areas within the high-rise flats—have hanging places for birdcages and an unused plot of land, the size of a football field, has been borrowed from the Government for holding song contests. Dotted with more than 200 vertical poles for suspending birdcages, it is a curious sight (see Colour Plate 18).

In 1985, a rough estimate put the number of songbird fanciers in Singapore at 25,000. An interest on this scale is more than just 'a relaxing hobby after a hard day's work'; it is a many-faceted one that has a powerful influence on the common man. Songbirds are an investment, they can build a reputation, enhance personal pride, bring people together, and provide pleasure and entertainment. Finally, what makes songbird-keeping really special in a Singapore context is that this is one of the few pastimes chosen by the people themselves, rising naturally out of shared interests.

4
The Most Popular Songbirds

ALTHOUGH there are many songbirds enjoyed by Singaporeans, there are only five that have become really popular: the Oriental White-eye, the China Thrush, the Red-whiskered Bulbul, the White-rumped Shama, and the Zebra Dove. Westerners have suggested that, of these, the Shama is the only true songbird, the others being only calling birds, but local people draw attention to the wealth and variety of the sounds they produce, and disagree. While the first four have some characteristics in common, the Zebra Dove is a different type of bird altogether, and for this reason is described separately.

These songbirds are fairly long-lived, especially in captivity, living up to ten or fifteen years. All of them are similar to the European starling or thrush in size (about 20–25 centimetres), except the White-eye which is significantly smaller (about 10 centimetres). Sexing them visually is rather difficult, although the mature male tends to be larger, with a broader head and a slightly brighter plumage. As females have no variety in their song, and are kept only for breeding, all caged songbirds are males.

The physical descriptions of the birds that follow are intended for the layman rather than the bird-watcher or bird-keeper, and have been deliberately made a very general guide. In each case, the shape of the bird's cage (described in Chapter 8) will also be a helpful clue to the identity of the occupant.

Oriental White-eye (*Zosterops palpebrosus*)

The Oriental White-eye is a tiny, active bird, with a slim, curved beak for taking nectar from flowers. It has a soft, olive-green body above, with touches of yellow and grey beneath. Locally, the bird is known by the Malay name of *Mata Puteh*, meaning 'White-eye', after the characteristic ring of white feathers around its eye (Colour Plate 7). These birds are indigenous to South-East Asia and groups of them are still found in the wild in Singapore, among coastal mangroves and bushy vegetation inland. They are no longer locally trapped, but are imported from Indonesia and West Malaysia. The most sought-after birds—*ayam* (chicken) birds, which can produce notes that rise to a strong staccato sound—come from Masai and Kota Tinggi in Johore, Malaysia. *Silu* (shy) birds, such as those from Taiwan or China, are not desired because they simply go 'chick chick'. These birds have no value because they are essentially marsh birds and their inherited voice is not one that is found attractive.

As a cage bird, White-eyes have rather a belligerent character, although they have never been used for fighting. They are probably the most difficult of the songbirds to keep because they are very sensitive to changes in the environment, and in their food and water. Drinking water from an area of Singapore other than what they are used to, can cause a drop in condition and even a premature moult. In spite of this, there may be more White-eyes kept in Singapore than any other type of bird, judging by the enormous number of entries in song contests. The Puteh, as it is usually called, is mainly a Singaporean interest; it is not widely enjoyed in Malaysia because there are restrictions on the number that can be kept and special permits are required. It is kept in Hong Kong and China, however, but has never been exceptionally popular.

China Thrush *(Garrulax canorus)*

The China Thrush has a dull, olive-brown plumage, with a white ring that encircles the eye and extends into a white streak behind the eye (Colour Plate 8). In Singapore, it is known mainly as the *Hwa Mei* (Cantonese, meaning 'beautiful eyebrow') or the *Weh Bee* (Hokkien). Among the non-Chinese it is simply called the 'Thrush'.

The bird is not native to Singapore and has been imported only since the 1960s, but already a number of escaped cage birds have established themselves in the leafy undergrowth of Singapore's parks and gardens. Meanwhile, wild birds from Guangxi in China reach Singapore through Hong Kong, and many are re-shipped to other countries in the region. Although thousands of Thrushes are exported from China each year, these still do not meet market demand.

As a cage bird, it is widely kept, but for different reasons. In Hong Kong and Taiwan, it is still used for fighting, and thousands of dollars are paid for the winners. Conducting fights between these birds was once a lucrative business in Singapore, too, with pieces of jade offered as the prize. Interest, however, shifted to its song, the shrill, repetitive nature of which prevents the keeping of many of these birds in the home. Enthusiasts often joke that their wives are driven to madness by the sound.

Red-whiskered or Red-eared Bulbul *(Pycnonotus jocosus)*

The Red-whiskered Bulbul is a greyish-brown bird with a white throat and white cheeks, and a patch of crimson under the tail and beneath each eye (Colour Plate 9). It is perhaps the most striking of the songbirds, distinctive for the shiny black crest on its head which moves up and down as it sings. The position of the crest is an indicator of mood, revealing whether the bird is bold or fearful. Locally, it is called by its Malay name of *Merbah Jambul*, meaning 'Crested Thrush', but it is usually referred to simply as 'Jambul'. 'Merbah' is often misspelled 'Merbok' (Jambul), but the true Merbok is a species of dove.

The Red-whiskered Bulbul is one of many species of bulbul indigenous to southern Asia. It is trapped wild in northern parts of Malaysia (Kedah, Penang, Pahang, and Kelantan) and Thailand (especially Bangkok and Chiang Mai), where it is plentiful. Most of those in Singapore have been imported from Thailand. Originally a forest dweller, the Jambul is now quite common in open country with some trees, and frequently found close to human settlements. Its appeal as a cage bird lies not only in its song, because this does not have a great deal of inherent variety, but in its lively nature and colourful appearance.

White-rumped Shama (Copsychus malabaricus)

The Shama or Shama Thrush, as it is known, even though it is not a true thrush, is the forest equivalent of the Magpie Robin to which it is closely related. Its local name is *Murai Hutan*, which means 'Forest Magpie'. Its body colours are sharply defined: bluish-black above with a white patch over the rump and a rich chestnut underside (Colour Plate 10). In Singapore, there are many more Shamas in cages than in the wild.

The Shama is a solitary, reclusive bird, always found in forested areas and often near water. It is widely distributed over southern Thailand, throughout the Malay Peninsula and the islands of Singapore, Borneo, Java, and Sumatra. The majority of competitive Shamas come to Singapore from West Malaysia and Penang and now increasingly from Indonesia and Thailand, depending on whether the purchaser cares more for appearance or for song.

There are various types of Shama, varying slightly in size, but the ones from Medan in Indonesia are the largest and most powerful singers. Shamas with long tails are found in Thailand and Indonesia and, to a lesser extent, in Malaysia, although Thai birds are often preferred for their more streamlined appearance. The most popular Shamas originate from the forested area along the Thai–Malaysian border. These birds are often comparable in size to the birds from Medan, but are more graceful and generally have longer tails.

Two-thirds of the male Shama is made up of tail, which it raises with a flourish while it sings. An impressively long tail is very much admired, especially if it can be held stiffly. The average length of a Shama's tail is about 19 centimetres, although it is said that recent disturbances in ecological balance have created a Shama with a shorter tail than in the past. While a good bird with a 28 centimetre tail is considered to be exceptional, one with a tail of 26 centimetres will fetch a couple of thousand Singapore dollars.

A disadvantage of the longer tail is that the usual 45–50 centimetre cage cannot accommodate it, necessitating a larger, more expensive cage of 60 centimetres. Also, the tail can be broken in transit or moulted out through stress, and may not grow back as long. Some unscrupulous people have been known to insert a wire pin in the hollow feather shaft and attach another piece of tail. Only when the cage is held up to light, does the dark patch on the tail reveal that two pieces of tail have been joined.

The Shama is famous throughout the world for its spectacular song. It also has an unusual ability to mimic sounds, from household noises to the sweet voice of the German Roller Canary. The female of the species sings as well, but its song is too short and erratic to be found pleasing. The male's impressive repertoire consists of three basic types of song accompanied by the appropriate body language: a loud song to establish territory, an aggressive song to defend territory and to dominate a female, and a gentle resting song or *tiong ow* (Hokkien for 'middle tone'), which can continue uninterrupted for hours on end.

The Magpie Robin (Copsychus saularis)

The Magpie Robin or Dayal Thrush, as it is also called, deserves a brief mention even though the Shama is now taking its place both in popularity and demand (Colour Plate 11). Its local name, *Murai*, means 'Magpie', but it was once referred to as the 'Straits Nightingale' on account of its melodious song. Its black body with white wings and white underparts was a very common sight in Singapore during the 1920s. Major Low, an army officer working in the Straits as a magistrate, states in his journal covering the years 1840 to 1841, that this bird 'may be tamed so as to require no cage, unless to protect it from the cats, and from the large rats which infest the straits ... he is so little afraid of man that he will sing for hours close to, or even on the top of a house ... ' (Moore and Moore, 1969: 202–3).

In 1987, fewer than twenty Magpie Robins were found to be left on the island due to their continuing popularity as a cage bird as well as the loss of much of Singapore's bushy landscape and mangrove thickets. Some are still imported from Peninsular Malaysia, although interest in its loud clear whistling, said to be too lacking in variety for use in a contest, has waned.

Zebra Dove (*Geopelia striata*)

The Zebra or Peaceful Dove is one of the smallest of the pigeon family, and is found in rural areas of Singapore and the Malay Archipelago. It is usually found in pairs or family groups in open country or around coconut groves. It is a shy bird, with a grey face and a pinkish-beige body marked with barred lines (Colour Plate 12). Locally, it is known as the *Merbuk* (Merbok), which simply means 'dove'. Doves are not strictly considered to be songsters, but both the Zebra Dove and the Spotted Dove (*Streptopelia chinensis*), or *Tekukor*, are prized among the Malays for their unique calls. Both of these birds have been an integral part of the domestic environment in the Malay Archipelago since ancient times.

The popularity of the Merbok as a cage bird spread from Indonesia to Thailand around the 1930s, where, from the 1950s, it was domesticated and bred. Later it spread to northern Malaysia (Kelantan) and to Singapore in the 1950s. The Indonesian Merbok was originally much sought after, but as the Merbok population declined from the late 1970s, most have been exported from Thailand to Indonesia and Singapore.

When it first became popular in Singapore, it was the *sandung* (or wild) bird from Indonesia that was kept. These are no longer popular and are considered of inferior quality. Their natural call, 'aou koh', has been gradually replaced with the loud rhythmic 'aou koh koh koh koh' of specially bred birds whose song has been finely tuned over the years. In the early days, few were bred and fine singers were not widely recognized. Both male and female sing, but the female, slighter in build with a narrower head and slimmer toes, does not have the stamina necessary to compete in a contest. They are quiet, unassuming birds and several can be kept in a single household, without their calls becoming offensive to the ear.

Although the Merbok is still kept mainly by Malays, and the words used to describe its behaviour are Malay words, it is also popular among Chinese and Indian owners, especially if they are Muslim. This is because of the belief that the Merbok imitates sounds from the Muslim holy book, the Koran. Continuing to find joy in the voice of this bird is synonymous with perpetuating the Malay culture.

5
Beginning with a Songbird

TRAPPING birds and plundering nests for their young has always been a popular sport in Asia, especially among young boys. In 1878, J. F. McNair found Malays to be 'very clever at catching birds by means of horse-hair nooses.... By this means doves and pigeons, some of them very beautiful, are readily taken; the juice of the gutta or indiarubber tree being sometimes used as birdlime[1] ... (McNair, 1878: 105).

The Merbok is still occasionally ensnared using this ground trap, although modern versions are made from a length of nylon string, strung between two sticks hidden in long grass. Several nooses hang from this, each with a special knot which allows the bird to be captured around the neck without hurting it. More frequently, birdlime of natural gum and latex is put onto a perch in the trees, using a decoy to lure the bird nearer. The mixture has to be strong enough to hold the trapped bird down, but fluid enough to be removed easily without damaging any feathers.

Henry Ridley, Director of the 'Singapore Botanic Gardens in the late nineteenth century, describes the process he witnessed as follows: '... the fowler conceals himself in a hut of leaves or ferns, provided with a cow's horn and a long stick with a loop of string at the end. Having sprinkled some rice on the ground in front of the hut he blows the horn so as to produce the cry "hoop hoop" of the pigeon. The birds come, and settling down

---

[1]Birdlime is a sticky substance traditionally used to immobilize birds.

before the hut begin to eat the rice, while the bird catcher nooses them one by one with the aid of the stick and string' (Bucknill and Chasen, 1927: 63).

For the other types of songbirds, a cage-like trap is set among the branches of trees, using a good singing bird as a *pemikat* or decoy. Early mornings are preferred for this, because birds are active and hungry at this time and will be easily tempted by a wriggling worm or insect pinned to a glue-laden perch. The Puteh is trapped in a small wooden cage with a piece of thread linking the perch to the open gate. When the bird ventures in, attracted by food or a decoy bird, the perch drops slightly, pulling the thread taut, and causing the gate to snap firmly shut. For Shamas, gates of soft netting prevent their long tails being damaged. Some traps have two storeys, with the lower storey containing the decoy bird. A typical Jambul trap might have a central area to hold the decoy bird, with compartments to the left and right, both of which might contain some juicy papaya to draw their victims nearer (Plate 8).

Trapping is now outlawed in much of the region, and although it continues in certain areas, trappers are fewer and wild birds are less plentiful. In addition, people have learned that even if they take great pains to travel to a popular area to trap a whole flock of birds and then separate all the males, there may not be more than a single bird that shows some promise of becoming a good songster. Captive breeding is thought to be the answer to this problem, the advantage being that young birds from fine parentage can be regarded as potentially good singers from the beginning.

Bird-trapping is illegal in Singapore, but can be enforced only if people are caught in the actual act of trapping. Traps are openly for sale in bird shops, as are

8. The Jambul trap in the foreground will keep the decoy bird in the centre section (which is so-shaped to accommodate the bird's tendency to jump upwards) while the Putch trap in the background will hold it in the lower storey. (Judith Balmer)

1. A Straits Chinese porcelain, or 'Nonyaware', plate, 11 cm in diameter, showing the phoenix, symbol of plenty. (Judith Balmer)

2. A Straits Chinese home in Emerald Hill, with the carved *pintu pagar* in the doorway, and blue-and-white tiles showing parrots in flight. (Lesley Layton)

3. Stamps from Singapore's 'Singing Birds' (1978) and 'Tourism' Series (1990). (Courtesy of the Singapore Philatelic Bureau.)

4. An Indian astrologer sitting by the roadside beckons to his parrot who will help to tell the fortune of passers-by. (Judith Balmer)

5. A Sunday morning at Petain Road, an old meeting-place for songbird lovers. A group of men are watching China Thrushes in the foreground while Mata Putch cages hang behind them. (Lesley Layton)

6. Tiong Bahru, the best-known coffee-shop, where proud owners meet to compare notes on bird-keeping and to hang their birds among others of the same species. (Judith Balmer)

7. An Oriental White-eye, or Mata Puteh (*Zosterops palpebrosus*) in a finely decorated cage. (Lesley Layton)

8. A China Thrush (*Garrulax canorus*) can be easily recognized by its long white eye-brow. (Lesley Layton)

9. Red-whiskered Bulbuls, or Jambuls (*Pycnonotus jocosus*) for sale outside a bird shop. (Lesley Layton)

10. A White-rumped Shama (*Copsychus malabaricus*). The angle of the bird's body and tail indicate that it is ready to begin its courtship song. (Lesley Layton)

11. A Magpie Robin, or Dayal Thrush (*Copsychus saularis*), famous for its glossy, pied plumage and melodious song. This bird is 'displaying', fanning out its tail feathers and opening its wings. (Lesley Layton)

12. A Zebra Dove or Merbok (*Geopelia striata*) in its cage. The cloth cover protects the bird from too much sunlight. The tear-shaped 'jewels' on the edges of the cage are a common adornment. (Judith Balmer)

13. A bathing cage has been placed against a Shama's cage (right) and the bird waits until it seems safe to dip into the water. (Judith Balmer)

14. Replenishing a bird's food is a daily necessity. A Puteh is being given a piece of orange while its owner enjoys a break at a coffee-shop. (Judith Balmer)

15. A man selling grasshoppers near the Tiong Bahru coffee-shop. (Judith Balmer)

16. An early riser 'walking' his birds. Note the beautifully carved and decorated Puteh cages and the size of the birds. (Judith Balmer)

17. Judges listening carefully in the last round of a Shama contest, and marking their scoresheets. (Lesley Layton)

18. A Merbok contest in progress, showing judges marking in their appointed areas. (Judith Balmer)

19. A fighting cage. Although fights are now illegal, the China Thrush and Magpie Robin are still housed in cages of this design. (Lesley Layton)

20. Unusually ornate Jambul cages indicate the growing popularity of this bird in Singapore. (Lesley Layton)

21. Thai influence on this Merbok cage is shown in the elaborately formed hook and other gold embellishment. (Lesley Layton)

22. Chinese influence on this Indonesian 'dome' cage is clearly shown in the painted lotus and the dragon feet. (Lesley Layton)

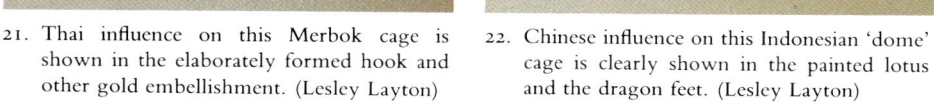

23. An early twentieth-century Chinese opium-bedside cage of bamboo and wood. Its accessories include blue-and-white feeders, ivory ornaments, a jade thumb ring, and a small swing to play on. The long legs of the cage enabled the bird to be easily seen when its owner reclined on the opium bed. (Judith Balmer)

24. An original set of antique (Ch'ing dynasty) porcelain birdfeeders of northern Chinese origin, showing a mountainous landscape hand-painted in enamels. (Judith Balmer)

the newly trapped birds, some smuggled across the causeway from Malaysia by car, others brought in by boat. Most are imported, huddled together in large rattan baskets or metal cages, distributed to bird shops and sold to the many bird lovers who cannot afford a songster with a fine pedigree (see Colour Plate 9).

Cages containing new birds are covered with a light cloth at first, and uncovered gradually over a period of a few weeks, before they are exposed in the shops. This reduces stress and prevents the birds from injuring themselves trying to escape. The best birds, caged separately for closer viewing, usually cost more, so some bargaining is allowed, whereas the more common types are inexpensive and bargaining is not expected.

Newly imported China Thrushes are usually housed separately because they are naturally aggressive. They are placed in small, self-contained wooden compartments with wire fronts, placed side by side in rows so that the birds are unable to see each other. Even a mature bird, unless being displayed or competing, is usually partially concealed under a light cover. This is to contain its aggressive instincts, as well as to prevent its strident voice from overpowering the song of another bird. Shamas are also housed separately in cages to protect their long tails from damage and because they always respond aggressively to birds of the same species, including females, outside the breeding season.

Choosing a good songster is a serious business which involves close observation of the birds, taking various points into consideration. Bird fanciers spend a great deal of time squatting in front of cages in bird shops, looking for some indication that the bird has the potential required. The importance attached to this choice is demonstrated by the shop that went so far as to place a plush red sofa outside for customers to make their decisions in comfort (Plate 9). A Singaporean breeder explained that 'for the ordinary man in Singapore, a Mercedes Benz is easier to come by than a really good bird'.

Potential buyers listen for a fine note, look carefully at the bird's physique, and register qualities of aggressiveness, stature, and plumage. It is said that the bird with a short beak is likely to make a good singer, while a long beak suggests a good fighter. In the past, owners of a fighting bird would have valued large, bold birds with strong legs and claws. Some say a young bird is best, an adaptable personality that has not become fixed in its habits, whereas others prefer a more mature bird that already has a song of its own. Males are always preferred because they sing as soon as they are born and (except in the case of the Merbok) their song can be improved with training.

Potential buyers will be concerned with details such as the manner in which a bird conducts itself in the cage. A Puteh may bend backwards and show a tendency to somersault, whereas a Thrush might have a habit of turning its head to look upwards. These are considered bad habits, even though they are not penalized during song contests.

Once a purchaser decides on a particular bird, he may have to keep sight of it among many others scrambling around the cage in fright. To catch a very timid Jambul, he will pick up a short cane stick with a noose attached, because it is said to be less stressful than a hand, slip this over the bird's head and draw it out of the cage. He may transfer a few birds to a separate cage in this way, to watch them closely before making a decision (Plate 10).

The discriminate fancier will search for a specific combination of physical attributes to ensure his purchase

9. The owner of this bird shop understands the importance of choosing the right bird, shown by the sofa placed directly in front of the 'new' Thrushes. A customer relaxes while the birdseller offers his advice. (Lesley Layton)

39

10. A group of men discuss the merits of some China Thrushes that have been moved to a wire cage for better viewing. (Lesley Layton)

comes closest to the ideal bird. He will choose carefully, taking into consideration the legs, the length of the bird's body, the size of its head, its eyes, the length and breadth of its tail, how the wings are held, the quality of the plumage, its posture, and overall health. For the Merbok, such physical attributes are not as important as its health and stamina, since its appearance is not judged in a contest.

Despite this painstaking attention to detail, many owners will later be dissatisfied with their purchase. Some birds turn out unexpectedly well, but at such an early stage there is no sure way to tell if the bird has the coveted qualities or not. There is perhaps a 60 per cent chance, or a greater one if it is from good parentage, but even sibling birds differ from each other.

The real value of each of these birds rests on the quality of its song. There is an interesting parallel between birdsong and human speech. Song is a bird's means of communication, and although the sounds are inborn, the form they take has to be learned from its parents and neighbours. The type of song is the same within any one species, although each bird does have its own individual voice, and those of the same species from different areas speak in different 'dialects'.

Birds' voices originate from a vocal organ called the syrinx, a powerful set of muscles which can manipulate air, setting off vibrations that produce a range of tones and pitches. In some birds these muscles can be strengthened, resulting in finer song. Although song is used mainly for marking and safeguarding territorial boundaries and for attracting females (especially among dull-coloured birds), there is evidence that the Shama sings purely for pleasure.

In addition to the quality of song, the birds must fulfil

certain conditions if they are to be trained to take part in a song contest. Therefore, every sound uttered is noted with care. For example, the Puteh's song is a sharp sound of two syllables, rapidly repeated, but in a contest the bird must also be seen to *buka*, which means to 'let go and bubble over with song'. The volume, tone, pitch, and melody of the song are also taken into account.

The China Thrush emits a hard, shrill series of 'tunes', all of the same pitch, with a periodic 'churring' sound. People gather to count the tunes, and the more they can count the more its value will rise. It is said that a good China Thrush can sing uncountable tunes. The Chinese say 'it is calling for its mother—it is calling for its nephew'—but it must not call for one of these too much without variation.

Enthusiasts of the Jambul enjoy its loud melodious whistles—four or five strong notes—which make up a 'tune' and can be quite high pitched. The bird will begin to attract attention if there are several variations of tunes and one spills naturally into another. For the Shama, a song worthy of a champion is one that is highly musical and continuously varied by the addition of jungle sounds. It should be loud and high pitched, include many rolling notes, and be able to last through four half-hour rounds of a song contest without fear of people, noise, or strange birds.

Throughout South-East Asia, the most important achievement for a Merbok is a series of 'steps' (intermittent phrases of notes which should be clear and evenly spaced), followed by a booming 'kong'. The Malays say that this sound should resemble the striking of the Malay *kompang* or hand-beaten drums. In Singapore, the aim is five steps or more, as well as the final 'kong', although four steps are more common and the 'kong' sound is not always clearly articulated. The prized vocal quality is present in very few birds, and to single out a potential champion from dozens requires a great deal of expertise.

The prices of these birds vary considerably. Newly caught wild birds are less expensive than those that have already been trained, or those that are captive-bred, and males always cost more than females. 'New' songsters are bought for a handful of dollars, but once trained they can become worth thousands of dollars, a price that can then double if the bird is sold in an expensive cage.

A trained Jambul and Shama will cost a few thousand dollars, but the price depends on which species happen to be most in demand at the time. One of the advantages of the Jambul is that it can be bought very cheaply, but after training can appreciate greatly in value. This makes them especially popular among lower-income groups, who can therefore regard them as an investment. Captive-bred Merboks are the most expensive, although even in 1946 Malays were 'said to pay many hundreds of dollars for one bird' (Loke Wan Tho Diaries). Today, a fledgeling could cost from tens of dollars to a few thousand, depending on its pedigree, whereas the offspring of a champion will cost several thousand, and may even be claimed by an enthusiast before it has hatched from the egg.

6
Songbird Care and Training

THE individual make-up of each species of bird dictates the kind of treatment it should be given. The Shama, Jambul, and Thrush are basically territorial birds, which are kept fairly separate from others of the same species, whereas the Puteh is gregarious by nature. These four birds have many aspects of care and training in common, while the Merbok does not require the same attention to detail. It is commonly referred to as the 'lazy man's bird', because it is so easy to look after.

Each of the songbirds has its own individual character. Some owners claim they would 'recognize their bird anywhere by the features, the shape of the body, the personality'. Contrary to popular belief, songbirds are capable of being affectionate, especially if raised from chicks. Kelham writes: 'My Malay *syce* had one which, on his approaching the cage, expressed its delight most demonstratively, fluttering its wings and cooing loudly, while a stranger made it wild with fear' (Bucknill and Chasen, 1927: 58). Such imprinting on man is no longer encouraged as these responses are inappropriate during song contests.

The birds are always hung high out of the reach of predators, close to a wall or in a corner for security, and away from draughts and direct sun. Sometimes they are covered with a cloth, either fully or partially, to rest them from singing, and from being stressed by other birds in the same house. Shamas, Jambuls, and Thrushes must be kept as far away as possible from other birds of the same species, or the continuous stress close proximity causes will wear them down until they cease to respond to each other. Conversely, Putehs need to be kept together in groups of three or more, otherwise they become 'solo birds', singing only when they are alone.

Songbirds must be bathed often to keep their plumage in good condition, because a restless, uneasy bird will not be able to sing continuously. There are special bathing cages for this purpose—small, rectangular, and roughly made, containing a metal trough about an inch deep which is filled with water. The Puteh, because it is so small, may have a small plastic or ceramic dish of water placed in the trough instead.

The bathing cage is placed against the birdcage, their gates pressed together, then opened (Colour Plate 13). The bird hops across the opening onto the perch above the trough and dips down into the water. The gates are then closed and the birdcage removed for cleaning. The base and the bamboo bars are brushed, several layers of newspaper lining are cut and fitted to the bottom, feeders are washed and dried, and fresh food is added. The birdcage is then put back against the bathing cage for the bird to return at will.

Young and newly trapped birds may return to their cage unbathed unless they can watch and copy an older bird. Some bathing cages have two or three compartments so new birds can bathe with others. The Merbok does not require a bathing cage because it is taken out of its cage and splashed with water gently in the hand. In the past, rice water—the water left over after washing rice—was used for bathing birds because it was thought to penetrate the feathers more easily than plain water. Some owners sun their birds for a short period daily after bathing, although the birds prefer the soft light of their natural habitat. The Merbok, whose cage top is always protected by a light fabric cover, can stay outside most of the day.

Before training begins, the bird must be placed on a good diet, because an undernourished bird will not stand a good chance of winning a competition (Colour Plate 14). Each species has slightly different demands: Shamas and Thrushes are basically insectivores, Putehs nectivores, Jambuls frugivores, and Merboks seed-eaters. A successful diet is one that comes as close as possible to that of the wild bird, and although some foods are enjoyed by more than one species in differing amounts, it is rare to find an owner that will give away the exact amounts and combinations of the ingredients he feeds to his birds. While most bird-keepers do experiment a little with diet, they are careful not to change it too often or too rapidly, because this may cause the bird to stop singing.

In the past, seed-eaters were fed on *padi* (unhusked rice), and fried, ground green beans; insectivores on earthworms, grasshoppers, cockroaches, and maggots from the fields; while local fruits were freely available for the others. Gradually, ingredients such as fish meal, chicken pellets, and alfalfa were added at home, and eventually the mixtures were sold in bird shops. Many of the songbirds are still fed on home-made mixtures but now these are more sophisticated, and may include chopped grasshopper, milk powder, chicken feed, wheatgerm and honey, vitamins A and E, and even dog food.

Fresh mixes may be more nutritious but commercial mixes are heavily relied upon for their convenience, even though they are expensive and the small packets do not last long. They contain special vitamin- and protein-rich ingredients, usually egg and bean mixtures, ground, fried, dried, and packaged locally. Many owners refry or bake the ingredients to reduce oil or moisture build-up and the possibility of rancidity, as well as to draw out the aroma. They may also add minced beef at this stage, but its oiliness is often blamed for spoiling the tone of a bird's voice. Overfeeding with rich food is thought to be one of the main causes of poor performance in a competition.

In the wild, the Merbok feeds on small seeds from various grasses and herbaceous plants. In captivity, its diet is *padi*, white and red millet, poppy seeds, and some herbs. Sometimes a special mixture such as spring onion or garlic pounded up with the leaf of the wild bitter gourd, is shaped into balls and fed to the bird to condition the voice.

Packaged food cannot make up for the lack of live food which most of the songsters need to maintain their song and keep them fit. Grasshoppers and crickets are the most common although dried waterflea, imported from China, may also be given to aid digestion. Meal-worms, too, provide an extremely rich source of protein, as long as they are given sparingly.

Grasshoppers raised on lush grass are imported from Malaysia in the dry season between monsoons. They are kept in mesh containers in the bird shops and sold by the hundred (Colour Plate 15). They are also sold at stalls during weekend song contests to 'heat up' the birds more quickly (Plate 11). The insects are put into a paper bag with some air holes and then transferred to small insect cages at home. They may also be placed in the refrigerator where they will 'sleep' for a week or so, becoming frisky again once they are removed from the cold. They are then fed live to the birds or, in the case of the very large

11. Business is brisk at a grasshopper stall during a weekend contest. The insects are taken home in brown paper bags. (Lesley Layton)

ones, speared onto small metal rods that are attached to the sides of the cages.

There are various grades of grasshopper, priced according to size. The grade of grasshopper bought depends on the needs and worth of the bird, though some varieties are considered too 'cooling' to be given at all. There is no relationship between the size of the grasshopper and the size of the bird. The largest grasshoppers are fed to the Puteh, slit open so that the bird can eat the soft parts inside, while the Shama will swallow smaller grasshoppers whole. Locally bred crickets are used when grasshoppers are unavailable, although they are not as nourishing. Cockroaches are known to be given for their medicinal properties, provided these are not urban insects that harbour harmful pollutants. It is more common for cockroaches to be used to teach newly caught wild Shamas to accept processed food, which sticks readily to pieces of the moist insect.

The majority of owners aim to enter their bird in a song contest one day. Once they have the songster of their choice, great care and effort go into the process of training the bird to sing. This may take up to a year and involve two main objectives: helping the bird to overcome fear, and improving its song. The bird must be capable of performing well in a crowded environment among unfamiliar birds. A well-trained bird will regard the presence of these as a trigger, and begin to sing almost immediately.

When the bird is young, it is encouraged to sing to itself for short periods each day, the cage partially covered so that it will not be easily distracted. This develops its natural voice before serious training begins, when the bird must be placed within the hearing, though not the sight, of one or more 'tutor' birds. This is especially important with Shamas, because they are very good mimics. Young birds are more easily trained, but the best age is from two to four years when they are more mature. Otherwise, unless they have a very dominant character, their song will be easily suppressed by an older bird. It may take even longer to bring the bird up to standard, and during that time it may even have changed hands a few times.

The traditional method of training is to take the birds outdoors regularly, in the early morning or evening of any day of the week, and especially on Sundays (see Colour Plates 5 and 6). Cages are hung together to provide an opportunity for the birds to listen to and sing against others of their species, the idea being that a concentration of songsters will borrow tunes from each other. They may learn new variations and pick up other sounds from wild birds in the trees nearby. They also get used to people, to being transported about, and to singing in a series of different environments.

The exact proximity of the cages is important. Two Shama cocks can stimulate each other to sing, and even complement each other's songs, but only if their cages are the correct distance apart. If any of the songbirds are placed next to another which is more dominant, they will cease to sing as frequently and to display, and need to be moved to another position (Plate 12). In spite of these attempts to avoid distress, the birds are being encouraged to out-sing each other and the experience puts them under enormous strain. For this reason, resting periods indoors have to be balanced against taking them out.

For people who cannot spare the time to bring their birds out often enough, a modern method of training is to play tapes to them of champion songsters of the same species. These tapes can be bought from most bird shops

12. A Jambul being moved to a better position. Owners must ensure that their bird is not intimidated by its neighbour during training sessions, especially if it is young or fearful. (Judith Balmer)

and are useful for new birds who need a stimulus to sing, for new owners to hear champion songsters, and to encourage birds that can imitate to vary their basic song. These tapes are also used during the birds' moulting period each year, to enable fine song to be absorbed immediately before the bird reaches its peak condition and enters a contest.

Some birds are inherently good imitators, but this will not by itself guarantee competition standard, because 'birds that just repeat a few tunes are dull, just as people who only answer yes or no when you ask them questions are dull'. The best singers are those that form interesting melodies. Training helps to increase the range and power of their song and this is what sets them above others.

The Puteh is the most demanding songbird because it needs to be 'walked' in company daily, or it will only sing when it is alone. Many of these birds are still owned by elderly and retired men who have the most time to devote to the bird (Colour Plate 16). The Thrush, Jambul, and Shama are more solitary and do not require to be brought out as often—perhaps once a week or once a fortnight. A short period of about half an hour daily is useful as a contest approaches, to build stamina without the danger of becoming too familiar with other birds.

The Merbok cannot be taught to imitate another bird's call regardless of environment or treatment. After the first year, its voice has already matured into a melodious sound that can continue for hours at a time. Training for this bird means being regularly exposed to other singers to build confidence and stamina. It must also grow accustomed to the height of the pole from which it will be hung during a contest.

The long period of training for all songbirds culminates

13. The National Songbird Contest, 1987, at the Jurong Bird Park. This is the Puteh section where entrants are beginning to assemble with their birds. (Lesley Layton)

in song contests, organized on a regular basis by community centres and residents' committees. A national contest, arranged periodically by the Jurong Bird Park in conjunction with The People's Association and the Tourist Promotion Board, attracts entrants from all over Singapore, as well as other ASEAN countries (Plate 13). Winning is a source of great pride.

7
Bird-singing as a Sport

IN the 1950s, 'playing birds' began to be taken seriously as a sport. Song contests were held for owners to test the ability of their songsters against others, and as a means of fund-raising for the community. In the 1950s and early 1960s, there were no formal, organized singing contests, just informal gatherings of enthusiasts at weekends and on public holidays.

The first informal contest was held in 1953 with only twenty entries. Formal contests began in 1964, after the start of the Kelab Burung Singapura, and represented four species of bird: the Thrush, Shama, Jambul, and Puteh. The Merbok was never included in these contests because it had little in common with the other birds and because the Malays were already 'very exacting as to the quality of the coo' of this bird (Bucknill and Chasen, 1927: 58), contests for the Merbok having started in Indonesia several years before they came to Singapore in 1959.

Since the early 1980s, bird-singing contests have attracted hundreds of entrants. Entry forms are filled in, sometimes a month in advance, and a small sum collected to cover the cost of awnings, trophies, and consolation prizes. Profits are always ploughed back into the community. There are also private contests which are essentially the same but may show small differences in organization and cost.

The birds' song is at its best during the breeding season, but contests are also held later in the year because captive birds come into season at varying times, and because the birds need the practice. Most contests begin in the early morning and end around noon. If it is a rainy day, there will be a noticeable drop in attendance because only seasoned singers will sing in damp weather.

On the appointed day, each name is registered and the entrant picks a number out of a bowl. This marks the place where the bird will be hung and cannot be altered unless an owner has two of his own birds next to each other. Each cage is hung a fixed distance away from the next and kept covered. The cover is removed only when the contest begins, so that the energy of each bird is released simultaneously. No bird can be moved once the contest has begun.

Not all of the birds begin singing with enthusiasm; some remain completely silent, feeding or preening themselves. For this reason, a judge spends the first ten minutes of the contest noting the birds' physique and display, before concentrating on the song (Colour Plate 17). A contest has four to five rounds and each round lasts half an hour. In the past, judges were scarce, but now there are judges for each species as well as those able to assess all four species. For each species, a team of four judges is used during the first two hours, one for each round. If there is a fifth round, the two most experienced judges mark half the birds each and the scores are averaged.

Points are awarded out of a total of 100: 80 per cent for song and 20 per cent for physique and display. Ideally, a judge is responsible for about thirty cages at any one time, but when the contests became really popular and judges were responsible for about sixty cages each, it became almost impossible to give each bird an individual mark. Points were then awarded from a fixed mark of 60, which meant that poor singers could be passed over quickly to concentrate on better ones.

Judges maintain that they have no need to look at the same bird more than twice. The mark that a pleasing singer deserves is noted on a scoresheet, never the maximum, but perhaps 75 per cent of this. If the bird turns out better still, then plus marks are added. A good judge does not need to add many pluses, and minuses are not permitted because this would upset participants. Marks are totalled in the 'counting room' and results are always on display to the public (Plate 14). Losing, it is said, does not mean that the bird was not able to sing, just that it was not in the right mood on that day. Winners, however, begin to establish a reputation.

In Europe, judges are expected to have a certificate of ability and a licence to assess songbirds in contests, but in Singapore experience and reputation count for much more. Information is passed on by word of mouth to newer judges who meet to compare notes after an event. There is no standardization, and bribery is an occasional problem. In some cases, a chief judge or overseer checks consistently high marks at each round to make sure they are genuine. Judges may either be paid, as in the case of Merbok contests, or receive souvenirs in return for their time.

Marks are noted on scoresheets according to the following criteria:

LOUDNESS (40 per cent)

Points are awarded according to the volume and clarity of the birdsong.
Puteh: Notes are joined quickly so they appear as a continuous flow of sound which can be surprisingly loud for the bird's size. The *buka* consists of a series of high-pitched sounds.

14. Scoresheets are displayed after each round and are eagerly read, even in the rain. Marks are jotted down on scraps of paper, or more often, on the palm of a hand. (Lesley Layton)

Jambul: Some of the more seasoned birds will *kekek* or cackle. This is a loud, aggressive, fighting sound which is only produced when the bird is in top form and trying to suppress other birds.

Thrush: Promising Thrushes can emit an ear-piercing melody.

Shama: Loud singers can be distinguished clearly even at quite a distance. Birds that have a very fine tone are at a disadvantage unless they can make themselves heard among louder birds.

VARIETY (20 per cent)

Points are awarded to birds with the ability to sing varied melodies.

Puteh: Many short, quick sounds are joined together to form a tune, but these are not always repeated in sequence. Although this is rare, as many as twenty varied tunes can be heard from this tiny bird.

Jambul: A variety of seven to ten short, staccato tunes are produced.

Thrush: Twelve to fifteen different tunes are not uncommon, with sudden changes in pitch.

Shama: This is the most important category for this bird. It can produce undulating phrases of notes linked together into songs, whistles, changes of tune and pitch, and dragged notes. These must be clear, natural sounds but can be punctuated with the sounds of other birds. If excited or alarmed, Shamas will also utter a sharp 'tock tock' sound.

STAMINA (20 per cent)

The stamina of birds is based on their ability to sing continuously or with very short breaks. A powerful bird that has the ability to drag six or more notes without a pause will be awarded the highest points. Stamina is judged per round, and the strongest of four rounds taken into account. Intermittent song is called 'breaking up', and does not show stamina.

Puteh and *Jambul*: As their songs are short, repetition of the same tune may occur, but other tunes must follow without too long a pause to maintain a continuing series of short songs.

Thrush and *Shama*: Continuous singing is expected, with only a short pause for display, or alternatively, singing and displaying at the same time.

POSTURE/DISPLAY (10 per cent)

Each species has different body movements which express its fighting spirit.

Puteh: This bird does not display, but a good posture is noted.

Jambul: The 'play' is more important in this bird than any other and marks are lost for its absence. The bird should exhibit 'jumping', or graceful rapid movements up and down the conical roof of the cage while singing. One without the other is unacceptable.

Thrush: Movements of the head, wings, body, and tail are taken into consideration. It should stand erect on its perch, neck outstretched, wings flapping and elevated, tail fanned and pulled downwards, and body feathers puffed to show boldness.

Shama: Points are awarded for body, tail, and wing movements. These variations in display take place within a series of movements: standing boldly upright on the perch while singing, stretching the head forward and down to perch level, bringing up the tail to its fullest height, and rapid movements from the perch to the edge or floor of the cage.

CONDITION/PHYSIQUE (10 per cent)

Points are awarded for health and condition. Bodies that are well proportioned are admired, and an unusually attractive bird may be awarded extra points, regardless of its song. For each bird the following attributes are noted:
Puteh: Sharp green and yellow feathers and round, bright eyes with thick, neat white rings.
Jambul: The crest pointed forwards on head, bright red-and-white feather markings on cheeks, and sharp pincer-shaped markings of the black feathers on the chest.
Thrush: An upright stance while singing, with outstretched neck and tight glossy feathers.
Shama: A long broad tail evenly and closely arranged, at least one and a half times its body length. A tail that is body length may not lose points against those birds with longer tails, although it is generally felt that there should be higher marks awarded for longer tails.

Merbok contests are also held in the early morning in residential areas and the procedure is much the same, except that the bird cages are drawn up by pulleys and cords to the top of steel poles 10 metres high (Plate 15). The height of the pole gives the bird a sense of security without which it will not sing well. Ideally, the poles should be 3.5 metres apart or the birds will find each other distracting; also the notes sung would be hard to distinguish.

The birds are divided into one of three categories, A, B, or C, depending on the pitch of their voices, and the field of poles is cordoned off into sections of about fifty

15. A Merbok owner prepares to raise his bird to the top of a numbered pole before the start of a contest. (Judith Balmer)

birds each. Entrants must leave the area as soon as the birdcages are hoisted so that judging can begin. There are four rounds, four judges that rotate from section to section, and each bird is judged on the best of three rounds (Colour Plate 18).

Scoresheets for the Merbok are small, with columns bearing the following headings, often abbreviated for convenience. Marks, out of a hundred, are not standardized but give a good average.

Angkatan: This means 'opening sound' and refers to the beginning or fore sound, 'How', which is a long, dragged-out sound.

Suara Tengah: This is the staccato middle sound, 'ke tek', which makes up the 'steps' in the Merbok's repertoire. A four or five step phrase is preferred.

Suara Hujung (Kong sound): The 'kong' is the end sound or finale. It can gain as much as 50 per cent of the points and make a difference to the price of the bird. All contest birds must have this 'kong' quality.

Although the 'How-ke-tek-tek-kong' sound may seem identical in each bird, both the middle and finale differ in strength and tone.

Irama: In addition to the above categories, each song differs in its melody, and enthusiasts appreciate the beauty of these differences besides the 'kong' at the end. The song must have individual flow and rhythm. Timing is important, too, for it must not flow so fast that it cannot be savoured.

Ayer Suara: The quality and clarity of the overall sound is assessed, which is the sum of all of the above categories.

Merbok contests are also popular in Malaysia and Thailand and many Singaporeans attend at least one of the three prestigious contests held in Thailand each year, which attract up to 2,000 entrants.

16. Trophies being prepared for presentation by a group of young scouts at the National Songbird Contest. (Lesley Layton)

Prizes for all the songbird contests consist of a series of large gilt trophies and smaller consolation prizes (Plate 16). Previously, trophies were small, flat, wooden wall plaques bearing an inscription, but to attract more participants they became larger and more ostentatious, until eventually a clock was incorporated into the design to make them practical as well as decorative. Trophy components are made in Malaysia and Japan and assembled later in Singapore, in various combinations. Prices depend on the number of components used.

The same birds often win again and again because their owners are very experienced and know their birds well. Each time a bird wins, its value rises. Champion birds sometimes become so expensive that it can be difficult to find a buyer for them. Gradually, their proficiency fades and they finally retire when they are about ten years old.

8
Cages and Accessories

IN South-East Asia, birds have traditionally been kept 'in an ingeniously made cage, formed of strips of bamboo arranged in a circle and bent over to a point, tied and furnished with a hook on the top' (McNair, 1878: 105). Such forms are still in use today although now they are rather more sophisticated and expensive. In Singapore, possibly more than anywhere else in the region, the cage is seen to reflect the worth of the bird.

The bird enters through a small, vertically sliding, and often decorated, gate of bamboo bars, to rest on a carved perch. The cage floor is held in place with metal clips and can be removed for cleaning. Below the floor of the cage, there is usually a metal handle for steadying the cage when it is being carried, while immediately above it, there may be a bamboo grille to prevent the bird's escape.

Only seasoned bamboo from China should be used to make birdcages because other types of bamboo snap easily when heated and bent into shape. A cage made entirely of bamboo is best, but because the material is scarcer now, cane is often used, mainly for the rim and base. Good teak and blackwood make fine, durable cages and usable pieces are often salvaged from old furniture for this purpose.

Cages are designed to attract attention and please the eye and there is always a demand for something different. This results in periodic fashions, such as the 'Ben Hur' cage that became popular in the 1960s and 1970s for the Merbok. The rattan band around the top was shaped like the wheels of the chariot seen in the film 'Ben Hur'.

Occasionally, a customer requests a cage in a specific design and has it custom-made in China, although traditional cages for each type of songbird are designed in a particular shape for a practical reason and with the bird's individual characteristics in mind.

A Puteh cage is a small, narrow, bell-shaped cage, the broad legs of which are carved with pictures and patterns which give the cage its name, for example, a 'prawn and crab' cage or even a 'teapot' cage (see Colour Plates 7 and 16). When it is bought, the interior of the cage is usually unadorned, but over time ivory feet, a tortoiseshell rim, or a decorated hook may be added. Its value rises with each addition.

A Thrush cage is a larger, bell-shaped fighting cage, recognized as such by its double gate (Colour Plate 8). When bird fights were common, these cages would be placed next to each other, and the inner gate of closely spaced bars raised so the birds could reach each other through the outer prongs. Magpie Robins are also traditionally kept in cages like these. (See also Colour Plates 11 and 19.)

A Jambul cage is square with a tall, pointed roof, shaped to accommodate the bird's acrobatic movements (Colour Plate 20; see also Plate 12). The base and four corners may be carved with patterns or auspicious symbols. The Jambul cage is said to have been designed in Singapore about forty years ago, based on a Thai prototype. Ideally it should be made of *jati* (pure teakwood).

A Shama cage is a large, cylindrical cage, up to 60 centimetres in diameter, to accommodate the bird's long and active tail (see Colour Plate 10). This makes it cumbersome to carry and often difficult to hang in a confined space, which may be why there are often fewer

Shamas than other songbirds entered in contests. The cage is usually simply decorated, it is said, so as not to detract from the beauty of the bird.

A Merbok cage is shaped to allow the cooing sound of the dove to carry better (Colour Plate 21). There are several variations of the basic shape, but the Indonesian version, the gaily painted 'dome' cage, is not very popular in Singapore, as it is said to artificially amplify the song of the bird. Striking carvings on its base and legs often show a strong Chinese influence (Colour Plate 22).

There is now very little cage-making carried out in Singapore. Most people are no longer interested in doing such intricate craftwork, particularly as many varieties of cage are imported and it is no longer cheaper to have one made locally. A few people still make cages for pleasure, and some have found it profitable, especially if their cages are well made and a little unusual.

Mr Chew, a shipbuilder by trade, began making birdcages when he was eight years old, after examining several cages to see how they were put together. He now regards his hobby as an art, for which he has to use imagination as well as accurate mathematical principles. In 1971, he designed a 'melon cage' for the Jambul after three years of experimentation (Plate 17). It is a rounded variation of the traditional square design and allows the bird to be seen more clearly. His respect for tradition, however, is shown in the odd number of bars that he always puts on the bottom of his Merbok cage for owners who may be superstitious.

After deciding on the size and design of the cage and making meticulous calculations, he heats the prepared body rings in the oven and puts them in a mould to cool. He then determines the position and spacing of the cage bars and ensures they are aligned with the rest of the

CAGES AND ACCESSORIES

17. The 'melon' cage, a new variation on the traditional design for the Jambul. (Lesley Layton)

cage. He cuts the base ring and glues it in a diagonal 'lap' join to minimize fatigue before wooden pegs are driven in. No nails are used, because they rust. He then shapes and sizes the cage bars and bends them to the desired angle by heating them over a flame (Plate 18). This is exacting work, because if they are held there a moment too long, the bamboo warps and burns. Ten bars are glued in to hold the body of the cage in shape before the rest are added. A wood dye is used to give the desired tone to the cage which, he says, 'should not look too bright in the light nor too dark in shadow'. Finally, a stainless steel hook and a carved perch are added.

Mr Amin Rahmat, a specialist in herbal medicines, is also a self-taught cage-maker. He makes traditional square Jambul cages in much the same way but he uses his instincts and the squares on his floor tiles as a guide to measurements. He has been innovative, too, making the cumbersome roof-end more graceful so that the cage appears lighter. He makes the base and then the frame, which he varnishes before the bars are added so that no area is left unprotected (Plate 19). To strengthen and steady the cage, he uses a strong glue as well as wooden pegs to join the roof to the corners where it meets the body.

Carving is done today using small electrical tools, rather than by hand. A design is traced onto wood or bamboo, and then pieces are chipped away with a chisel and knife or fretsaw, so that the image stands out in relief (Plate 20). Wooden cages are best for carving human faces or intricate flowers. Bamboo splits easily and is generally confined to simple surface patterns.

In the past, a customer would choose the design to be carved on his cage from a book of drawings. Malay and Chinese motifs are different and are rarely done by anyone outside the appropriate ethnic group, although one

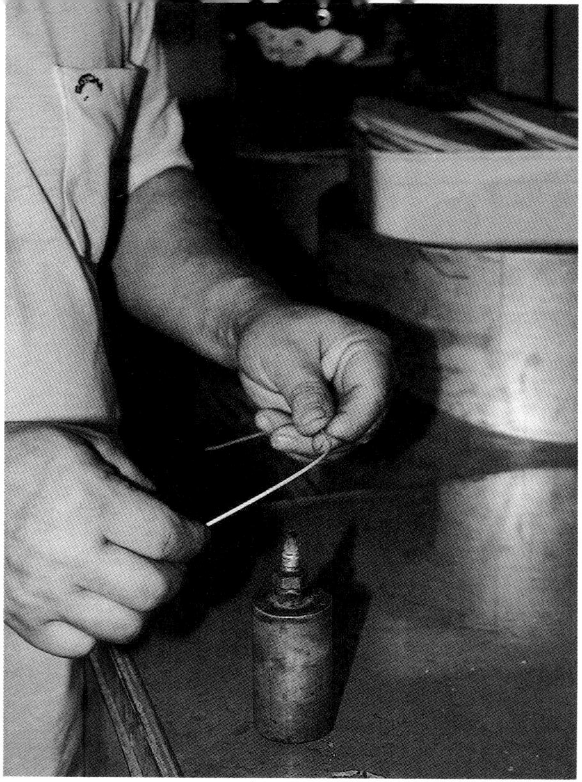

19. An unfinished Jambul cage. The bars have yet to be added and varnished, after which a hook and a carved perch will complete the cage. (Judith Balmer)

18. Individual bamboo bars for the cage are bent gradually as they are heated over a flame. (Lesley Layton)

old Malay man, Pak (Uncle) Molok, is fondly remembered for his faithful and inimitable rendering of Chinese designs onto cages costing a few thousand dollars.

Birdcages are often imported unvarnished because good quality varnish is unobtainable in China. Shellac may be used as a coating to prevent the cage warping, before a layer of varnish or lacquer is applied, because humidity turns bamboo brittle. Varnish is popular because it can be sprayed on quickly and covers up flaws in cheap cages. Lacquer has to be applied more carefully by brush but it is a good waterproofing agent. In the kampongs of the past, the leaves of a wild creeper called *memplas* were used to finish cages, smoothing rough edges and giving a dull polish. Alternatively, coconut

20. (Top) A drawing is lightly pasted onto the section to be carved. In this case, holes are drilled right through the wood before being shaped into an open fretwork design. (Bottom) A finished product, showing a typical Malay design carved in ebony. (Judith Balmer)

leaves burned to a powder would be rubbed on and dusted off and then a tree wax used for a richer colour and gloss.

Unvarnished cages are often imported unassembled. The top is packed inside the base with the cage bars pre-bent and bound with string. These are sent to a cage-maker or repairer who will assemble, varnish, and return the completed cage, even adding a bit of carving if desired.

Many people own cages and there is a steady income to be made from cage repair. Bamboo is fragile and areas commonly in need of attention are gates, bars, and corners which are easily knocked and damaged. A broken bar is coated with thinner to loosen the varnish before it is removed. The new one is cut to size, inserted, and heated to bend it into place. Where there is major damage, such as a vertical break in the body ring, it is better to buy a new cage.

Every cage buyer is looking for something beautiful, but most of all he is looking for something unique that other people will admire. Achieving this may cost a lot of money. The overall price of a cage will depend on its age, the quality of craftsmanship, and the materials used, but can cost hundreds of thousands of dollars. For cages no longer in production, collectors might pay more than the cage is actually worth, desiring it for its own sake rather than because they are bird-keepers.

Small decorative items are believed to add class to the cage, increase its value, and draw attention to the bird, which may be for sale if the price is right. These might include an antique perch of black coral for wearing down the bird's toenails, a tray for waterflea in old ivory, or a jade thumb-ring, a safe and fashionable way of carrying cages in the nineteenth century (Colour Plate 23).

Puteh cages are almost always ornamented with boxwood or ivory carvings that are traditionally auspicious: flowers, small creatures, or miniature opera scenes, usually in pairs. The Chinese are very superstitious, but today symbols such as these are to reinforce pride in the cage rather than in any true belief that they will bring luck.

Some additions to the cage are practical as well as decorative, such as the tiny cages, 5–10 centimetres square, of bamboo, tortoiseshell, or ivory, which are clipped on the inside or outside of the cage and contain live grasshoppers for the birds to peck at. Cage covers, often of batik, in medium dark natural colours (Plate 21), are used to calm the birds when they are being transported, frequently by motor bike (Plate 22). Covers for Merbok cages are made in Thailand, and the Thai influence is often clearly shown in their bright colours and use of gold embellishment. (See Colour Plate 21.)

Among cage accessories, perhaps the finest are the old Chinese birdfeeders, the earliest of which were earthenware and date back to the T'ang (AD 618–906) and Yüan (AD 1271–1368) dynasties. During the Ch'ing dynasty (AD 1644–1911), they were made in larger quantities, mostly of porcelain, although other materials include carved and uncarved wood, precious materials, such as ivory or jade, as well as glass and metal.

Antique cage accessories are found mainly in private collections, though some bird-keepers use them to decorate their finest cages. These mainly comprise feed and water bowls in sets of three, four, or even six pieces (Colour Plate 24), but also include insect trays, bathing trays, and flower vases. Flower vases were originally meant to be hung on the outside of the cage, with a flower placed inside, to sweeten any odour from the

21. The cage cover calms the birds when they are being transported and batik is a popular choice of material. Most covers zip up one side so the bird can also be partially uncovered when necessary. (Judith Balmer)

22. In Singapore, as in other areas of the region, birds are frequently transported by motor bike. (Lesley Layton)

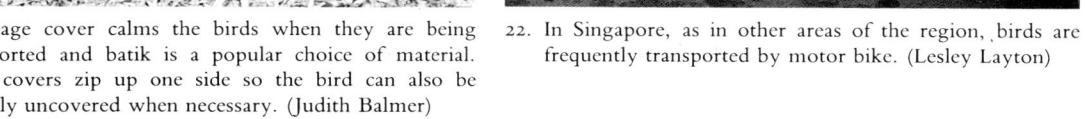

23. Two elderly men having a light-hearted discussion over a pair of modern birdfeeders. (Judith Balmer)

bottom of the cage. Complete sets are preferred because they are more valuable, but rare individual pieces are also prized. There are many different sizes and shapes, depending on the function for which they were intended, as well as on the size of the bird. Rough porcelain copies of these old designs were made from porcelain in Johore until finer ones became available just as cheaply from China (Plate 23).

9
The Future of the Songbird Tradition

THE rapid growth of Singapore in the 1880s signalled the beginning of an alarming decline in the wild bird population. In *The Singapore Naturalist* in 1923, F. N. Chasen, Curator of the Raffles Museum at the time, wrote: 'The gradual extension of the city must perforce drive the birds away ... a few years ago green pigeons were to be seen in the Raffles museum compound and kingfishers flew up and down the canal in Stamford Road. Such events are now remarkable' (Vol. 1, No. 2, April 1923). This trend has continued to the present day.

The swamps and jungles have gone, the number of tall trees has diminished, and high-rise buildings have multiplied. While the Jambul has been able to adapt to these enormous changes in the environment, other birds that were once so common, such as the Magpie Robin, have become very scarce. Their disappearance is due not only to the erosion of their natural habitat, but to trapping on a very large scale.

Statistics show that although many wild birds have been lost over the years, some are returning. Singapore's gardens, patches of open land, and bird sanctuaries have attracted several species, and their movements are carefully monitored. Laws have been passed to reduce the capture and trading of wild birds, and citizens are becoming increasingly aware of the need for such constraints.

Meanwhile, the interest in captive songbirds has grown rapidly. Unlike the local wild birds of the 1960s that were weak, plump, and capable of only a limited range of songs, their successors were stronger and their song better developed. This was because, over the years, owners singled out specific attributes, and restricted themselves to birds from certain areas. As time went by, fewer 'good' birds became available, people had to travel further to find them, and permits became necessary to move them from one country to another.

Owners found themselves expecting a great deal more from their birds than most wild birds could offer. Impressive-looking birds that could learn to produce loud, sustained, and melodious songs were scarce in the wild, but these were the birds with the most value. As a result of the belief that such fine attributes are inherited, there came a general shift in focus towards captive breeding. In Thailand, the venture has been so successful that whole villages have become involved in the breeding of the Merbok.

In Singapore, only the Merbok is bred commercially, the other songbirds being bred mainly as a hobby by a few experienced owners. Breeding on a large scale for most interested people is hampered by the shortage of space and the high cost, not only of building aviaries and accumulating good stock, but also the difficulties in obtaining the large amounts of live food required daily by the parent birds and their young. The Shama becomes totally insectivorous immediately after the chicks are hatched as the parents will not feed processed food to the chicks.

The Puteh, Jambul, and Thrush are imported into Singapore in large numbers and are relatively inexpensive. To invest a great deal of time and money into breeding them locally would not be economically worthwhile. Occasionally they may be bred to further the strain of a particular champion bird, or by someone who wishes to learn more about the species. The Shama attracts greater interest, and among breeders there are two main schools: those who try to reproduce conditions in the wild as much as possible, and those who encourage the wild bird to adapt to a smaller, more restricted environment.

Putehs do breed easily and rapidly once they are given nesting materials (Plate 24), as do Jambuls, even on an apartment balcony with a healthy pair selected from a bird shop. They must have a home that provides a sense of security and is fairly undisturbed, but most important of all is a sufficient supply of live food for the chicks to mature, after which they will accept other foods. Breeding the Thrush is rare in Singapore, because almost all imported birds are males. Putehs, Jambuls, Thrushes, and Shamas lay two to four eggs after an incubation period of about ten days to two weeks. Wise breeders do not allow

24. A man-made Puteh nest used at the Jurong Bird Park to encourage these birds to breed. (Lesley Layton)

these birds more than two clutches of eggs each year, or the size of eggs per clutch tends to become smaller, and the rate of infertility increases.

A specific pattern of behaviour, common to each of these songbirds, is exhibited in the wild, and has a direct bearing on the duration and quality of their song. The wild baby bird acquires its adult plumage between six and ten months. After this first moult, it begins to take on the behaviour of the mature bird. It is quiet and timid before and during the nesting period. It then becomes more hostile, and its song gradually alters in frequency and volume, until it reaches its loudest and most forceful during the breeding season. Although at this time a mated pair respond aggressively towards any trespasser within their territorial boundary, they will return to their previous quiet state when the breeding season is over (Figure 1(a)).

This natural cycle alters in captivity. A protected indoor environment, with constant supplies of the same food, causes physical changes to occur at a different pace than they would in the wild. Caged birds may take longer to moult, and mature several months later than they would in the wild. Alternatively, a premature moult can be triggered by stress. Reaching 'peak condition' in the period of the breeding season when the bird is in full song, highly excitable, aggressive, and responsive, varies therefore from bird to bird, and even the same bird can vary from year to year. Three birds in the same household can reach this stage at different times within the same breeding season (Figure 1(b)).

Of all the songbirds, the Shama is the only one to exhibit a dramatic sign of having reached this stage. The roof of its mouth changes from a pale red to a red that is

THE FUTURE OF THE SONGBIRD TRADITION

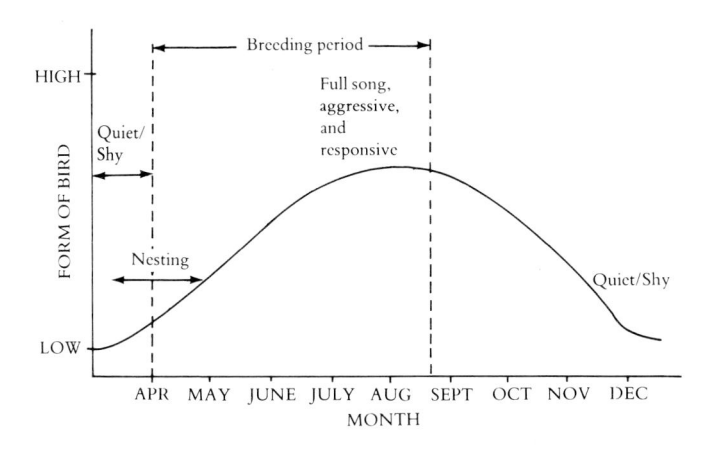

Figure 1(a) The Shama's behaviour in the wild follows a natural cycle which is constant and predictable.

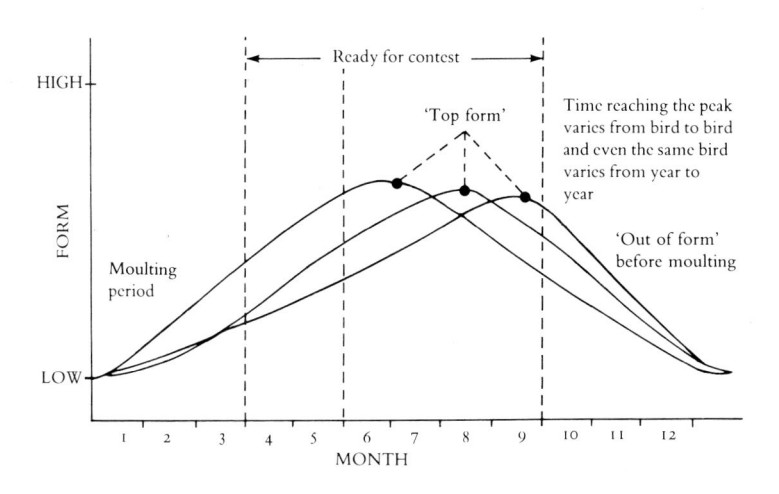

Figure 1(b) In captivity, the Shama's natural behaviour varies with changes in the bird's care and environment. (Figures courtesy of Shama breeder, Mr Lim Heau Dong.)

almost black. This is when its future potential as a competition bird can be more accurately predicted, and is the best time to enter it in a song contest. The birds are considered 'out of form' in the period between the end of the breeding season and the next moult, the following year.

Breeding requires an owner who understands these changes, and is sensitive to his birds' needs. Timing is crucial and untimely judgements can spell disaster. A male Shama, introduced to a female at the wrong time, may kill her. Keeping a close eye on the baby birds, and intervening if and when it is necessary, is a painstaking task. Many songbird-keepers are men with full-time jobs who care single-handedly for several birds, often of different species. Some feed them on home-made mixtures and occasionally try to breed, not only the birds, but also the mealworms or grasshoppers to feed them. Most admit that there are not enough hours in the day to care for their birds as they would wish, and find breeding a full-time job in itself.

One of the pleasures of breeding is demonstrated by the experienced breeder who found, contrary to popular belief, that the song of his locally bred Shama surpassed the song of the wild Shama in richness and variety. As Shamas are fiercely territorial, the young in the wild hear the song of other Shamas only from a distance and from time to time. Whilst they will imitate the songs of some other species of bird in the vicinity, they will never be exposed to the variety of birdsong to which the really serious Shama breeder can expose his chicks.

Merboks are bred at breeding farms in outlying areas of Singapore, their simple diet, and the ease with which they can be kept, proving to be a great advantage. They are bred all the year round in captivity, although there are

THE FUTURE OF THE SONGBIRD TRADITION

concerns that this is causing the offspring to become smaller with time. Enthusiasts prefer to buy birds from such places, however, because the chances of acquiring a good bird are greater when the parent birds are known. The offspring are sold locally, as well as exported to Indonesia. Outside Singapore, finer pedigree songsters are raised in Malaysia and the southern provinces of Thailand.

In Thailand, the demand is so high that the young are often booked in advance. Champions are paired with choice females and their offspring are inbred. Two eggs are laid in one nesting, and apart from occasional three-monthly rest periods, the female lays repeatedly. This is because the eggs are removed as soon as they are laid, and placed in an incubator. When they hatch, the chicks are placed with 'foster' birds that care for the young. Later, when they are sold, the leg ring that they bear confirms their origin. The Malays often name a bird based on its lineage, such as *bapa ribut* (literally, 'storm father'), whose offspring will then be *anak ribut* ('storm child').

A typical breeding farm will hold about 500 pairs of birds and a healthy pair, in secure surroundings with ample food, can produce ten broods in twelve months. Unlike the other songbirds, the quality of the Merbok's song remains constant all the year round. Moreover, the lavish care and feeding accelerates the development of the chicks, resulting in younger birds winning contests. For these reasons, breeding Merbok is a lucrative business.

And what of the future? The practice of 'playing birds' is not expected to die out. Younger people are beginning to get involved and, although it is still a predominantly male passion, some women are taking an interest. Certain practices deviate in small ways from accepted methods within the rest of the region, but the hobby is still young, and Singaporeans are finding their own individual expression of it.

The immense popularity of the hobby indicates that there is much more to it than appears on the surface. In a country where self-expression is often not encouraged, it offers people opportunities to be individual; where different races make up a single nation, here is something they all have in common; where much is borrowed from the West and provided ready-made, 'playing birds' is truly theirs—inherited, deep-rooted, and familiar.

When asked what they felt about the future of their pastime, and of the songbird tradition in general, a cross-section of Singaporeans had the following to say:

Grasshopper seller: (On the ancient practice of keeping Merbok for their song) 'It cannot stop, it is too old now.'

Experienced breeder: 'People here still don't know what they are doing. When each fad comes along they follow it, learning by default, rather than having a true and tested tradition to follow.'

Businessman: 'It can never die. It is an interest that lasts a lifetime and is often passed on from father to son.'

Housewife: 'If they (referring to husbands in general) are bird-keepers and they give up their hobby at some point, they have only to hear another [song]bird, and they will buy one again.'

Bird shop owner: 'Nowhere else in the world is there a hobby where so many people of different race, class, and financial status can get together with a common purpose.'

Old man in his seventies: 'If they offer me three thousand, I won't sell the bird. If I sell it, I won't have a good bird to "play" anymore.'

THE FUTURE OF THE SONGBIRD TRADITION

Cage-maker: 'I don't think the songbird tradition will ever die out. The competitions may well do so, people may get tired of those, but for the person who keeps the bird for himself, for its sound, and pleasure in the sound, it can never die.'

Select Bibliography

Bird, Isabella, L., *The Golden Chersonese and the Way Thither*, London, John Murray, 1883; reprinted Kuala Lumpur, Oxford University Press, 1967, and Singapore, Oxford University Press, 1990.

Braddell, Roland, *The Lights of Singapore*, London, Methuen & Co. Ltd., 1934; reprinted Kuala Lumpur, Oxford University Press, 1982.

Bucknill, A. S. and Chasen, F. N., *Birds of Singapore and South-East Asia*, Singapore, Government Printing Office, 1927; reprinted Singapore, Graham Brash, 1990.

Loke Wan Tho, The Personal Diaries of, entry under 25 March 1946, National Archives, Singapore.

McNair, J. F., *Perak and the Malays: Sarong and Keris*, London, Tinsley Bros., 1878; reprinted Kuala Lumpur, Oxford University Press, 1972.

Moore, Donald and Moore, Joanna, *The First 150 Years of Singapore*, Singapore, Donald Moore Press, 1969.

Tate, D. J. M., *The Chinese Empire*, Hong Kong, John Nicholson Ltd., 1988.

———, *Straits Affairs*, Hong Kong, John Nicholson Ltd., 1989.

Wise, Michael, *Travellers' Tales of Old Singapore*, Singapore, Times Books International, 1985.

Index

References in italics refer to Colour Plate numbers.

Animal market, 13

Bathing, 45–6; cage for, 45, *13*
Beliefs and superstitions, 4, 7, 19–22, 68, 74
Bird market, 9
Bird ownership: as pets, 7, 13; in early Singapore, 13–15, 19; *see also* Ethnic groups
Bird shops, 12, 13, 15, 23, 35, 39, 40, 50, 79
Bird symbols: in architecture, 3, 4; in art, 1, 8; on ceramics, *1*; in religion, 1, 4, 7, 8
Birdfeeders, 74, 76, *24*
Birdseller: Chinese, *2*; Malay, 7, 9, *10*
Breeding, in captivity, 32, 35, 43, 78–83; season, 55, 80–2
Bucknill, A. S., 21; *see also* Chasen, F. N.
Budgerigar, 14
Bulbul, Red-whiskered (Jambul), 14, 15, 26, 28–9, 35, 36, 38, 42, 43, 44, 46, 51, 52, 55, 59, 60, 61, 67, 68, 69, 70, 71, 77, 78, 79, *3, 9, 20*

Cage accessories, 73–6; birdfeeders, 74, 76, *23, 24*; cage covers, 46, 74, 75, *21*
Cage cleaning, 45
Cage design: for fighting, 67, *11, 19*; importance of, 26, 66, 73; modern, 66, 69; traditional, 66, 67–8, *7, 8, 10, 12, 20*
Cage-making, 68–73; cost, 43, 73; finishing, 71, 72–3; materials, 1, 66, 67; repair, 73
Canary, 14, 15, 31; German Roller, 31
Carving of cages, 70, 72, 73; motifs, 67, 68, 70, *7, 16*
Champion birds, 42, 43, 50, 52, 65, 78, 83
Chasen, F. N., 21, 77; *see also* Bucknill, A. S.
Cockatoo, 7, 12, 14, 15, 19; *see also* Parrot
Coffee-shops, as meeting places, 23–4, *5, 6*
Cormorant, 5
Crane, 4

Dayal Thrush, 31; *see also* Magpie Robin
Decoy bird, 35, 36
Diet, 15, 46–9, 78, 82, *14*
'Display', 50, 60, *10, 11*
Dove, 16; Peaceful, 32; Pink-headed, 15; Spotted, 15, 32; Zebra, 7, 14, 15, 19, 20, 21, 25, 26, 32, 33, 34, 38, 41, 42, 43, 44, 45, 46, 47, 52, 55, 57, 61, 62, 63, 68, 74, 78, 82, 83, 84, *12, 18, 21*
Duck, 4

Eagle, 3
Environment (domestic): in the archipelago, 32; in housing estates, 15, 17, 24–5; in kampongs, 15, 16, 19, 72–3
Ethnic groups, viii; Chinese, vii, viii,

INDEX

1, 4–5, 12, 14–15, 19, 20, 22, 74; Eurasian, 15; Indian, viii, 7–8, 15, 19, 22; Malay, viii, 7, 9, 12, 15, 19, 20, 21, 22, 32, 33, 43, 55, 83; (Straits) Chinese, 4, 15, *1*, *2*

Fighting birds, 5, 6, 19, 24, 28, 38, *8*, *11*; fights, 22, 23; *see also* Cage design

Goose, 4
Grasshopper, 47–9, *15*; cage, 74; *see also* Diet

Hanging out of cages: frequency of, 52; places, 23, 24, 25, 50, 67, *5*, *6*, *18*; positioning, 37, 44–5, 50, 56, 61, 62
Hornbill, 7
Hwa mei, *see* Thrush, China

Jambul, *see* Bulbul, Red-whiskered
Java Sparrow, 5, 15

Kelab Burung Singapura, 23, 55; *see also* Singapore Caged Birds Society
Kingfisher, 77
'Kong' sound, 21, 42, 63

Lark, Mongolian, 14; Sky, 14
Legislation regarding birds, 22, 24, 35, 77
Lorikeet, 14

Meeting places (for bird lovers), 23–5, *5*, *6*; *see also* Trade in birds
Merbuk (Merbok), *see* Dove, Zebra; *see also* Song contests
Moulting period, 52, 80, 81; premature moult, 27, 30, 80
Murai, 31; *see also* Robin, Magpie
Murai Hutan, 29; *see also* Shama, White-rumped
Mynah, 14, 15, 19; Hill, 14

Owl, 7

Parakeet, 12, 19; Hanging, 14, 15; Ring-necked, 15, *4*
Parrot, 5, 7, 8, 12, 13, 19; *see also* Cockatoo
Peacock, 4, 8
Pheasant, 4
Phoenix, 4, *1*
Pigeon, 5, 7, 34; Green, 77
'Playing birds', viii, 55, 83, 84
Purchasing songbirds, 37–41; attributes, 21, 38, 61, 78; cost, 43, 65; selection, 38, 39, 40; undesirable traits, 21, 38
Puteh, *see* White-eye, Oriental

Quail, 5, 6, 7, 22

Robin, Magpie, 13, 14, 21, 22, 31, 77, *11*; Pekin, 15

Shama, White-rumped, 14, 26, 29, 30, 31, 35, 37, 41, 42, 43, 44, 46, 49, 50, 52, 55, 59, 60, 61, 67, 68, 78, 79, 80, 81, 82, *3*, *10*, *13*, *17*
Singapore Caged Birds Society, 13, 23; *see also* Kelab Burung Singapura; Meeting places
Song: link with human speech, 41; of individual species: Jambul, 29, 42; Magpie Robin, 31, 32; Merbok, 32–3, 42; Puteh, 27, 42; Shama, 31, 41, 42; Thrush, 28, 42; organs for, 41; *see also* Song contests
Song contests, 8, 18, 25, 44, 47, 49, 53, 54, 55–6, 68, 82; criteria, 57, 59–61, 63; judging, 56–7, 63, *17*, *18*; procedure, 55–6, 61; scoresheets, 57, 58, 63; *see also* Breeding season
Songbirds: affection from, 44; age, 26, 35, 38, 49–50, 65; as cage birds, 27,

INDEX

28, 29, 32; 'display', 50, 60, *10*, *11*; females, 26, 31, 33, 37, 43, 82; local names of, 27, 28, 29, 31, 32, 83; 'new', 35, 37, 39, 43, 45, 52; outside Singapore, 27, 28, 29, 30, 32, 42, 55, 63, 78, 83; physical appearance and characteristics, 26–33, 41, 44; species, *see* individual species

Taming, 7, 31
Thrush, China, 14, 15, 22, 26, 28, 37, 38, 39, 40, 42, 44, 46, 52, 55, 59, 60, 61, 67, 78, 79, *3*, *5*, *8*, *19*
Trade in birds, 9, 12, 14, 28, 77

Trained birds, 5, 24, 41, 43; training process, 49–52
Trapping, 14, 19, 27, 29, 34, 35, 45; as a sport, 34; traps for, 5, 34, 35
Trophies, 64–5

White-eye, Oriental (Puteh), 14, 23, 24, 26, 27, 35, 36, 38, 42, 44, 45, 46, 49, 52, 53, 55, 57, 59, 60, 61, 67, 74, 78, 79, *3*, *5*, *7*, *14*, *16*

Zebra Dove, *see* Dove, Zebra
Zoo, birds in, 13

Sophie decides to go on a diet (though she can't stop eating chocolate and chips!). She feels fed up because she hasn't got a boyfriend ... but there's a boy on a bike who seems very special.

Will Sophie ever get to see him again?

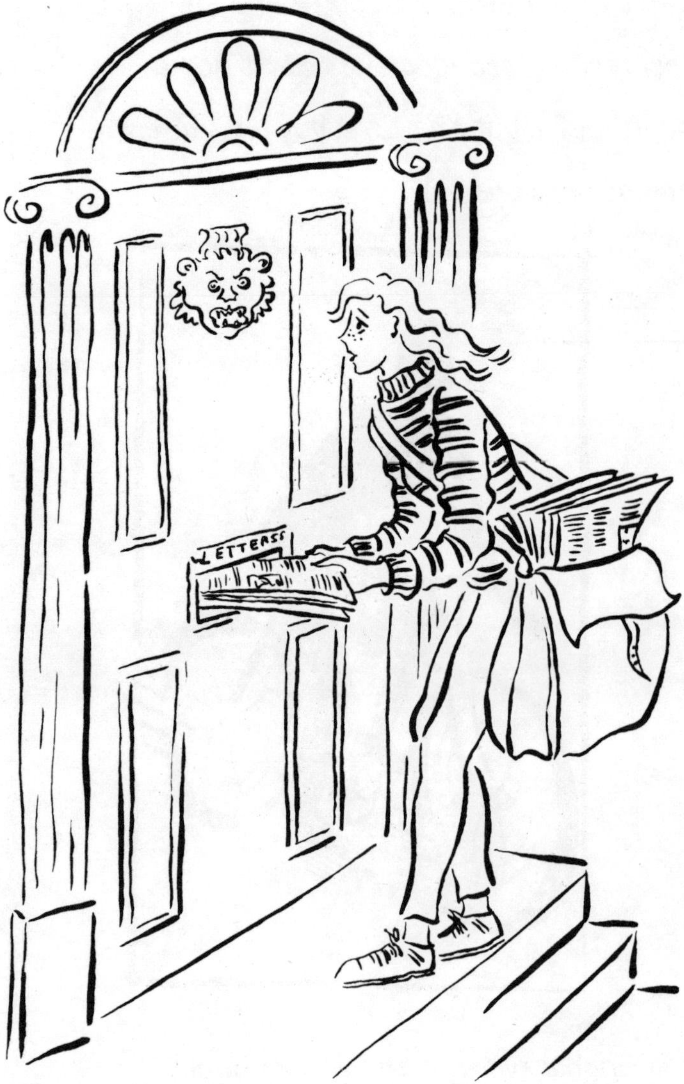
LETTERS

SUNDAY

What do you do on Sundays?

I bet you have a long lie-in. I have to get up at six o'clock. Yes, six!

I have a quick dab in the bathroom, pull on a jumper and leggings, and that's me, fit to face the day. I *don't* think!

My dad doesn't look too great either. He's downstairs, sorting the newspapers. My dad's a newsagent. So guess what? I have to be the newspaper girl.

"I keep looking for a kid to do the rounds, Sophie," said Dad. "But kids nowadays don't seem to want a newspaper round."

You're telling me they don't. Especially not this kid.

"It's not fair," I said. "Why can't Becky or Sharon take a turn?"

I know why. My sister Becky's dead brainy and she's set on getting all these A grades in her exams. She stays up half the night doing her homework. So *she* can't get up at six.

My sister Sharon's not a bit brainy, but she's ever so pretty and popular. She stays out half the night with all her pals. So *she* can't get up at six.

7

I'm not like Becky. I'm not like Sharon. I'm the stupid stay-at-home sister. So *I* have to get up at six. And it's not fair! I do the paper round during the week. That's not so bad. I do it on Saturdays too. But the Sunday round is a killer.

Every newspaper has all these special magazines. They weigh a ton. I can hardly lift the bag. I don't know who my dad thinks I am. Arnold Schwarzenegger?

I fell off my bike this morning. I went over a stone and went all wobbly. The bag slid to one side. Over I went, clonk! I must have looked a right idiot, lying there with my legs in the air.

This boy came along on his bike and looked round at me. I felt myself going red.

"Are you OK?" he said.

"I'm not sure," I said.

"Here," he said, and held out his hand.

He picked me up. Then he picked my bike up. Then he picked my bag up too. "Hey, it isn't half heavy!" he said.

"You're telling me," I said.

Then we seemed to run out of things to say. We just stood there.

"Well. Better go. See you," he said.

"See you," I said.

I wonder if I *will* see him? What's he doing out on his bike at this time of the morning anyway? He had a bag too, but it wasn't big enough for newspapers. Lucky him.

MONDAY

Monday. School. Yuck.

You'll never believe this. My sister Becky likes school!

We go to school on the bus together. I tried to get her to do my homework for me. "Go on, Becky, please. You know I'm hopeless at sums," I said.

But she wouldn't.

"Cheer up, Sophie," said Sharon. She comes on the bus too, though she's dead lucky and left school last year. She works in the supermarket.

"It's daft, all that stuff they teach you in school. Sums and that," said Sharon. "You don't need it when you get a job. You just work the check-out and it adds it all up for you."

Sharon didn't get A grades in her exams. Or Bs. Or Cs. It doesn't matter though. Sharon likes her job.

And she's the sort of girl who looks great in a naff nylon overall. She makes us laugh telling us about all these shoppers who keep chatting her up at the check-out.

"Sharon," I said on the bus. "How do you get a boy to chat you up?"

"Ooooh," said Sharon. "Hear that, Becky? Our Sophie's getting interested in boys. Who is it, eh?"

"Yes, who is it, Sophie?" said Becky.

I tapped my nose. My sisters can be dead nosy at times. There isn't anyone, anyway. Not really. Apart from the boy on the bike.

TUESDAY

I've been looking out for him on my newspaper round. I haven't seen him yet.

I should have done my maths homework in break, but I read this magazine instead. My friend Carol and I had a chuckle at the problem page letters at the back.

Then we started reading about this new diet. It seemed like a great idea. So we went on it right that minute.

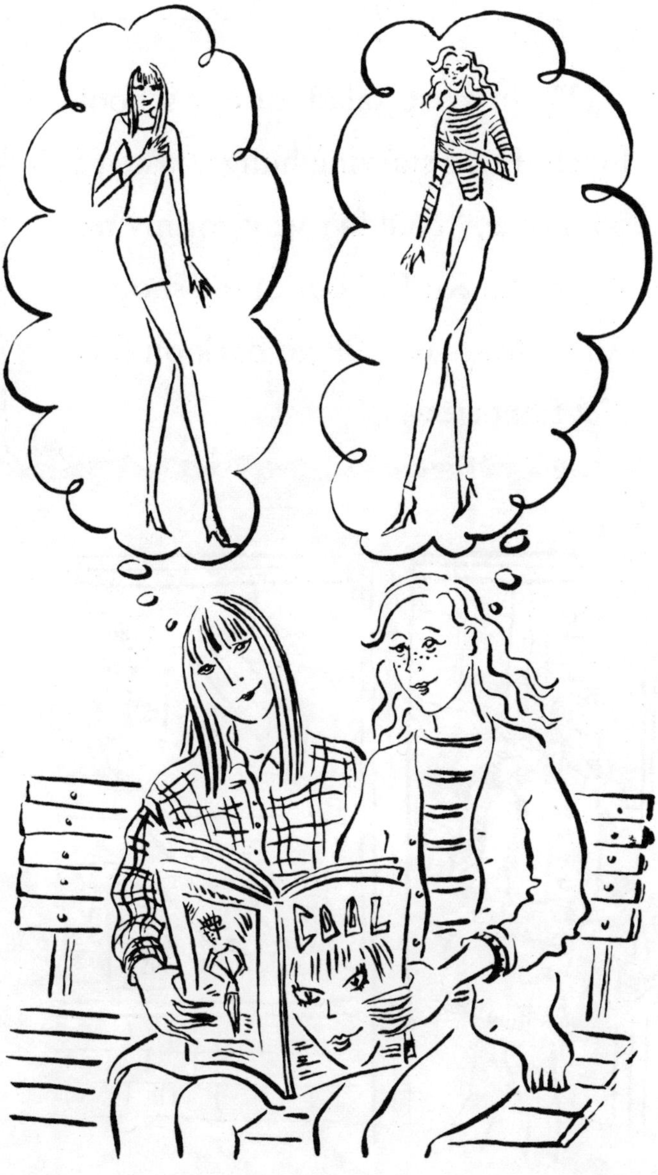
COOL

We just had salad for our school lunch. I was starving hungry when I got home. I couldn't wait for my tea so I had a Kit Kat out of the shop. And a Mars Bar. I'll go back on the diet tomorrow.

WEDNESDAY

I tried not to have any breakfast but Mum said I'll make myself ill. Especially as I have to get up so early.

"I wish we could get someone to do the round," she said. "Poor old Sophie. You need a proper lie-in. You're looking pale and peaky."

I sneaked into Sharon's room and put a lot of her blusher on. Just in case I might see anyone on my way to school.

Carol and I had salad again at lunch but we got a bit fed up. The girls at our table all had chips. So we had chips too. And then we had doughnuts. Yummy, yummy, yummy.

We can always go on a diet tomorrow. And I needed cheering up because that mean old Miss Marker made me stay in after school. Because I never did my maths homework.

But guess what?!!!

I saw the boy on the bike on my way home and he waved to me.

I didn't even realise who it was at first. Then I went bright red. I didn't need Sharon's blusher.

I didn't start waving back until

he'd gone past me.

I waved and waved and waved.

THURSDAY

Carol was feeling dead depressed this morning. She thinks she's getting fat.

She isn't really. Apart from her bottom. I wouldn't *say* that, of course. Carol's my friend.

Carol kept going on about it in P.E. I told her it didn't look big at all. Though our P.E. shorts don't help!

Mrs Bright, the P.E. teacher, came bouncing up to us. "You're on this court to play netball, not stand there nattering," she said. "Run about a bit! Get some exercise. That's the way to lose weight."

I hate exercise.

Mrs Bright runs all over the place. And you should see the size of her bottom!

FRIDAY

I felt dead embarrassed in assembly. My sister Becky is a prefect. She had to read in front of the whole school. She did it really well. But all my friends were giggling.

It's odd having such a swot for a sister. Carol didn't giggle. She was looking pale and peaky. Then she suddenly slid to the floor.

Becky had to stop reading while we carted Carol outside. We thought she might be really ill but she'd only fainted. Because she hadn't eaten any tea last night. She hadn't eaten any breakfast either.

Mrs Bright gave her ever such a talking-to. So Carol had fish and chips for lunch. So did I.

Then we went to this café on the way home and had ice-creams. Lots of boys hang out there, especially on Fridays.

It was my idea to go to this café. Just in case the boy on the bike goes there. He didn't go there this Friday. But this other boy started chatting to Carol. She started chatting back.

I started to feel like maybe I was in the way. So I went home. Feeling a bit depressed.

Carol rang me later. Guess what! She's going out with this boy tomorrow. So Carol's got a boyfriend before me. I felt even more depressed.

"Cheer up, Sophie," said Dad. "Great news! A boy popped into the shop this afternoon wanting a newspaper round. He's called David. He's starting tomorrow. So you don't have to do it any more."

"You can have a lie-in at last," said Mum.

SATURDAY

I had a lie-in.

It was great. Only maybe not as great as I'd hoped. I still felt dead depressed when I got up.

"Cheer up, Sophie," said Becky. She seemed in a very good mood. "Look, are you still having problems with your maths? I'll go over it all with you, shall I?"

"Thanks – but no thanks!" I said. I was feeling depressed enough without the maths.

Mum doesn't work on Saturday.

"Cheer up, Sophie," said Mum. "Look, come shopping with me."

"No, I'm meeting Carol. I'm going shopping with her," I said.

Only that wasn't a good idea either. We went round and round the shopping centre looking for a new outfit for Carol to wear tonight. When she goes out with this boy.

I don't need a new outfit. Because I'm not going out with a boy.

Yes I am!

Yes I am, yes I am, yes I am!

You will never ever guess what.

It happened this evening. I was feeling dead dead dead depressed.

Sharon was hopping about in her high heels ready to go out. Then Becky put all her books away. She started getting ready too.

"Where are you off to, Becky?" said Mum.

"I'm going to the pictures," said Becky. "There's a French film on. It'll be good for my French."

"Who are you going with, Becky?" said Dad. Becky suddenly looked like she'd been at Sharon's blusher too.

"Well ... David, actually." David, our new newspaper boy.

Only he's not really a boy. He's at the sixth-form college. He needs to earn some spare cash so he can go out at nights.

So he can go out with our Becky!

So that was everyone in the whole world going out except me. Even Mum and Dad decided to have an evening out down at the club.

We keep the shop open late on a Saturday. I minded it while Dad had his tea and Mum did her hair.

I was feeling so depressed that the customers kept saying, "Cheer up, Sophie."

But I didn't feel like cheering up, did I? Until ...

Until the door opened and in came the boy. The boy on the bike. Well, he didn't ride his bike into the shop, obviously.

He left the bike outside. He came in. He saw me. I saw him.

I went bright red.

He went bright red too.

"Hello!" he said.

"Hello!" I said.

Then we just sort of looked at each other. And smiled.

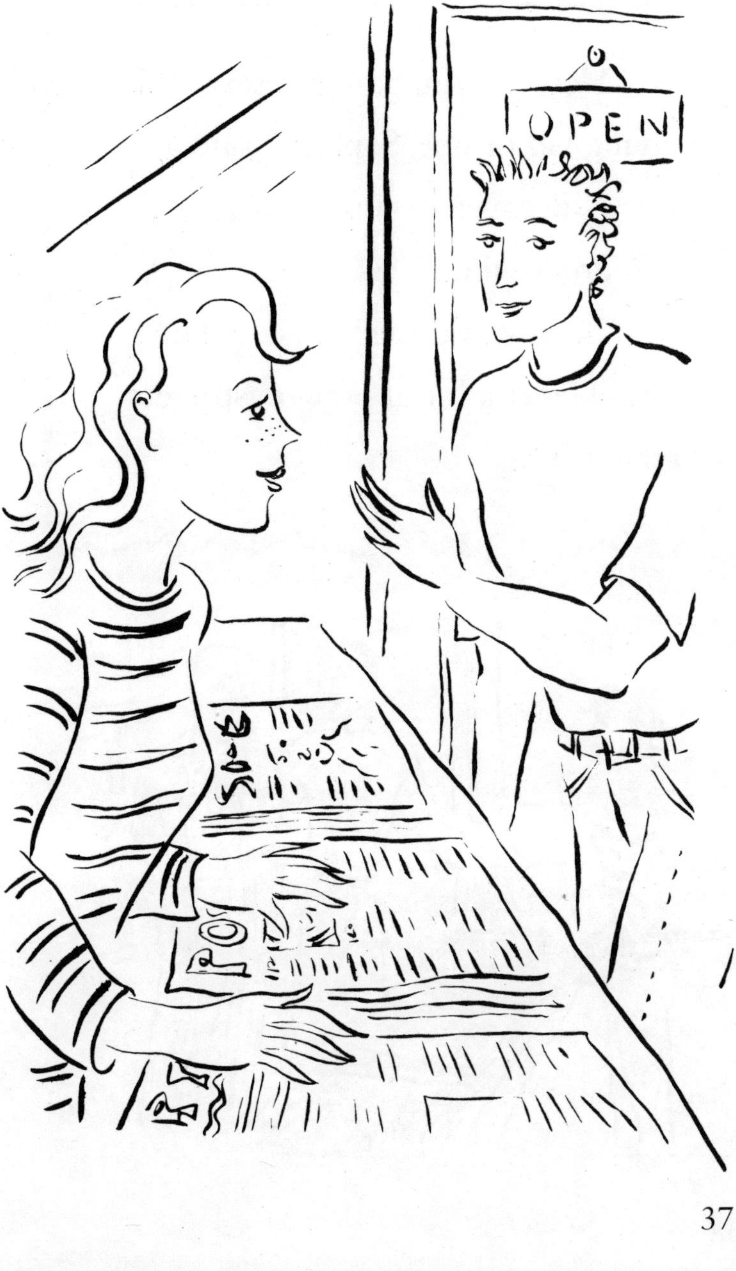
OPEN

"Hey, are you going to serve this young lad or not, Sophie?" said a man. "Because I want my paper, if you don't mind!"

The boy said the man could go first. He let a lot of other customers go first too.

He stayed in the shop chatting to me.

Well, we didn't do much chatting. But it was OK.

"So your name's Sophie, right? That's a pretty name," said the boy.

"What's your name?"

"Robbie."

"Oh really? I like that name a lot."

Then I served some more customers. "I've been looking out for you, Sophie," said Robbie.

"Have you?" I said.

"Yes, since last Sunday, when you fell off your bike. I saw you one time after school. But I don't think you saw me."

"Yes, I did."

"Oh! Well. Anyway. I was sort of wondering ..."

"Yes?"

"Whether you'd like to go out with me sometime?"

"Oh! Well. Mmmm. Yes."

"Really? Great! Only I can't go out tonight because my nan's come round on a visit. Oh! That's why I'm here. To buy her some chocolates. But tomorrow night?"

"Yes. Tomorrow night's fine. Now, what sort of chocolates does your nan like?"

We chose a box together.

Then my dad came in to shut up the shop.

"I'd better be off," said Robbie.

"Yes," I said. "I hope your nan likes her chocolates."

"Can I just buy this little chocolate heart too?" said Robbie.

"Sure," said my dad, taking the money.

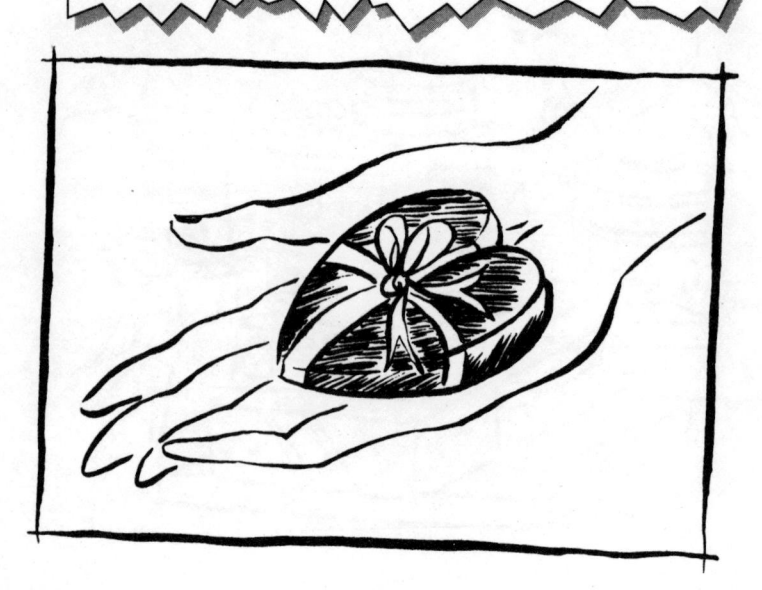

The chocolate heart wasn't for Robbie's nan.

He gave it to me!

I felt dead embarrassed actually, in front of my dad.

But thrilled too.

"See you tomorrow evening then," said Robbie.

"Right," I said.

"It seems ages away. Hey, how about seeing me in the morning first? I go to early-morning swimming on Sundays. Do you fancy going swimming? Oh no, wait, you've got your paper round."

"No, I haven't!" I said. "Not any more. I'd love to go swimming with you, Robbie."

"So it's a date, right? I'll call for you at half-past six. Don't you mind getting up early?" said Robbie.

"Of course I don't mind," I said.

My dad was shaking his head. "I

can't believe my ears!" he said, when
Robbie went out of the shop.
"You're going swimming, Sophie?"

"It's about time I started to get a bit of exercise," I said.

"At half-past six in the morning? I thought you were longing for a lie-in?" said Dad.

But he was only teasing. Mum teased too, when he told her.

And Becky. And Sharon.

I don't care.

There's only one thing I care about now.

One boy.

Robbie.

Oh well. I'd better stop writing now. I've got to get up early in the morning.